Lecture Notes in Physics

354

J. De Coninck F. Dunlop (Eds.)

Wetting Phenomena

Proceedings of a Workshop on Wetting Phenomena
Held at the University of Mons, Belgium
October 17–19, 1988

Springer-Verlag
Berlin Heidelberg GmbH

Editors

Joël De Coninck
Université de Mons, Faculté des Sciences
19, avenue Maistriau, B-7000 Mons, Belgium

François Dunlop
C.N.R.S., Ecole Polytechnique, Centre de Physique Théorique
F-91128 Palaiseau Cédex, France

ISBN 978-3-662-13793-2 ISBN 978-3-540-46966-7 (eBook)
DOI 10.1007/978-3-540-46966-7

Originally published by Springer-Verlag Berlin Heidelberg New York in 1990
Softcover reprint of the hardcover 1st edition 1990

The contributions in this book were processed by the authors using the $\mathrm{T_EX}$ macro package from Springer-Verlag Berlin Heidelberg GmbH.

2153/3140-543210 – Printed on acid-free paper

Foreword

Our department of Statistical Mechanics and Probability is very pleased with the organisation, for the second time, of a workshop on wetting phenomena. Wetting is a very lively subject in statistical physics and we shall surely have the opportunity of hearing of how the many aspects and tools of this branch of physics apply to the field.

May I also rejoice in the international participation in this workshop.

Mons,
October 1988

Ph. de Gottal

Preface

This volume contains the proceedings of the workshop on wetting phenomena held in the University of Mons in October 1988.

Many problems of practical importance involve the wetting of a solid by a liquid or more generally the wetting of a surface dividing two phases by a third phase (paint, lubricant, aerosol, metal coating, etc). The study of these phenomena at a microscopic scale has, however, been developed only recently, giving new insight on dynamical interface profiles, layering transitions, multiphase wetting, influence of disorder, adsorption, and so on. The mathematics of random surfaces has also become a very active field of research.

The need for sophisticated tools, both experimental and theoretical, and the interdisciplinary nature of the subject, have made it both useful and sometimes difficult to have meetings between the different specialists. The Mons workshop and these proceedings bring together samples of recent advances from many different approaches to wetting.

We thank the contributors for making their lectures so appealing that work extending beyond the interfaces of this fascinating subject is bound to have been stimulated.

The financial support from the Région Wallonne and the Université de Mons is gratefully acknowledged.

We also thank Marie-Anne Carlier for assistance in preparing the workshop and these proceedings.

Mons,
October 1988

Joël De Coninck
François Dunlop

Contents

MAGNETIC WETTING TRANSITION

J.C.Bacri, R.Perzynski, D.Salin

Laboratoire d'Ultrasons[1], Université Pierre et Marie Curie
Tour 13, 4 place Jussieu
75252 PARIS CEDEX 05, FRANCE

Abstract

We follow the spreading of a magnetic liquid, a non wetting ferrofluid, along a wire. The external control parameter of the spreading length of the fluid is the magnetic field generated by a current travelling through the conducting wire. The spreading length results from a balance between capillary and magnetic forces. For a current threshold, we observe a rapid jump of this length corresponding to a sheath-like coverage of the wire. This magnetic wetting transition is analogous to the wetting transition on a fiber, predicted for a totally wetting fluid in the presence of van der Waals forces. The longer range of magnetic forces leads to a transition at a macroscopic scale. The crossover between micro and macro scales is studied.

1 Introduction

If the main features of the wetting of a liquid on a solid has been clarified two centuries ago, in the pioneering works of Young and Laplace [1], it is only recently that much effort has been paid to understand basic capillary problems such as controlled parameters of wettability [2], the wetting-dewetting transition [3], and the dynamics of spreading [4,5,6]. Let us become more precise on the simple problem of the ascension of a fluid along a vertical plane : the ascension height results from a balance between gravity and capillary forces and is of the order of the capillary length (a few millimeters for classical fluid and solid). In fact this description of the interface is macroscopic and is correct in case of partial wetting of the solid by the liquid, i.e.

---

[1] associated with the Centre National de la Recherche Scientifique

when there is a finite contact angle θ between solid and liquid (if $\theta > \pi/2$ the ascension is a depletion). For complete wetting ($\theta = 0$) the macroscopic picture is still valid but with a microscopic precursor film first observed by Hardy [7] ; the typical thickness of a few nanometers of this film results from a balance between gravity and attractive van der Waals forces. Of particular interest is the understanding of the spreading of this film under external parameters such as temperature for binary mixtures [3,8] or magnetic field in case of magnetic fluids [9].

In this article, we describe the magnetic wetting transition of a magnetic liquid, a ferrofluid [10], on a wire. The spreading is controlled by the magnetic field generated by the current travelling through the wire.

2 Wetting of a Fiber with a van der Waals Fluid

Compared to the wetting of planes, the wetting of fibers exhibits some particular features due to the additional curvature [11]. Let us imagine a fiber of radius b covered with a liquid film of thickness e0 (we restrict our discussion to the case of non volatile liquids) as represented in Figure 1.

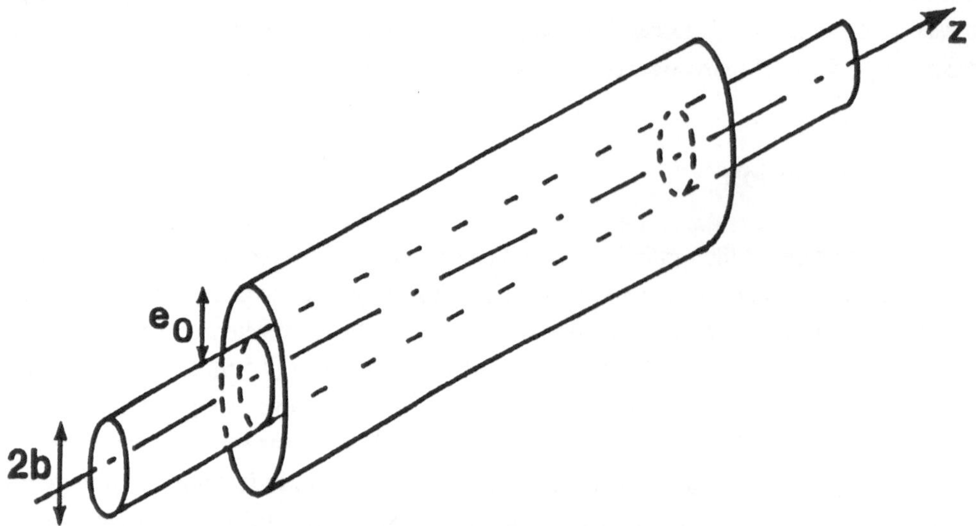

Figure 1 : A fiber of radius b covered by a liquid of thickness e_0 .

The interfacial energy W (per unit length) of such a situation compared to the dry fiber case is :

$$W = 2\pi b\,(\gamma_{SL} - \gamma_{SO}) + 2\pi\,(b + e_0)\,\gamma$$

where γ_{SL} , γ_{SO} and γ are respectively the solid-liquid, solid-air and liquid-air interfacial tensions. Introducing the so-called spreading parameter [5] :

$$S = \gamma_{SO} - \gamma_{SL} - \gamma\,,$$

we get :

$$W = 2\pi b\,\left(- S + e_0 \frac{\gamma}{b}\right) \qquad (1)$$

We immediately see that to get a complete wetting situation, we should have

$$S > e_0 \frac{\gamma}{b} > 0 \quad ,$$

so that we may imagine a system (solid/liquid) for which the liquid will wet the plane (S > 0) but not the fiber !

Let us start with the description of van der Waals interactions which anyway exist for every system in nature [5,12]. If we consider a film of thickness e, its van der Waals energy per unit area is (in the limit of non-retardated interactions) :

$$P(e) = -\frac{A_{SLO}}{12\,\pi e^2}$$

where A_{SLO} is a Hamaker constant associated to solid (S) and air (O) separated by a liquid slab (L) :

$$A_{SLO} \approx \left(\sqrt{A_{LL}} - \sqrt{A_{SS}}\right)\left(\sqrt{A_{LL}} - \sqrt{A_{OO}}\right) \approx \left(\alpha_L - \alpha_S\right) . \,\alpha_L$$

where the α's are the polarisabilities of solid and liquid. For complete wetting conditions, $\alpha_S > \alpha_L$ and thus P(e) > 0 (we assume here that the interfacial tensions are due to dispersion forces). The energy (1) becomes :

$$W = 2\pi b\left(- S + e\frac{\gamma}{b} + \frac{\gamma a^2}{2e^2}\right) \qquad (2)$$

with a defined by

$$a^2 = -\frac{A_{SL0}}{6\pi\gamma}$$

(a is a molecular size, typically a = 3Å). Minimizing W with respect to e leads to :

$$e = ec = (a^2\, b)^{1/3}$$

and defines a critical spreading parameter S_c [11]

$$S_c = \frac{3}{2}\,\gamma\left(\frac{a}{b}\right)^{2/3}$$

For $S < S_c$, we have a non wetting (≡ partial wetting) situation. The liquid remains unspreaded and forms a drop on the fiber with a contact angle given by the Young-Dupré relation

$$\gamma \cos\theta = \gamma_{S0} - \gamma_{SL}.$$

For $S = S_c$, the thickness of the wetting film at the threshold is

$$e = e_c = (a^2 b)^{1/3}\ .$$

For $S > S_c$, the thickness is

$$e_0 = -\frac{A_{SL0}}{4\pi S}$$

(as for plane surfaces [13]). If the liquid is "in excess", i.e. the volume is larger than

$$(\pi(b+e_0)^2 - \pi b^2).L$$

where L is the length of the fiber, the final situation will be a reservoir drop in equilibrium with a wetting film of thickness e_c.

Magnetic interactions can also form wetting films and they have the advantage to be of longer range $(b+e)^{-1}$ instead of e^{-3} per volume unit without any divergence in the vicinity of $e = 0$) and, moreover, they are easy to be monitored in amplitude.

3 Wetting of a fiber with a magnetic fluid

Besides the experimental difficulty to observe wetting at microscopic length scales involved in van der Waals interactions (< a few tens of Å), it would be of interest to monitor, through an external parameter, the transition from a non wetting to a wetting situation. Such a wetting transition has been proposed to interpret experiments in a binary mixture close to the phase separation temperature with the temperature as an external parameter [8]. Once again the observation scale is very small. Thus the use of a magnetic fluid with the magnetic field as an external parameter is suitable for wetting studies. Following the same approach as above but for a conducting wire of radius b through which a current of intensity I is travelling, we get for the energy per unit length W :

$$W = 2\pi b\left(-S + \gamma\frac{e}{b} + \gamma\frac{a^2}{2e^2} - \gamma\frac{L_H}{b}\mathrm{Ln}\left(1 + \frac{e}{b}\right)\right) \tag{3}$$

The additional magnetic term introduces a magnetic length

$$L_H = \mu_0\chi\frac{I^2}{8\pi^2\gamma}$$

μ_0 is the vacuum permeability and χ the magnetic susceptibility of the ferrofluid. Minimization of W together with the condition of energy decrease ($W \leq 0$) leads, for a given S value, to a film thickness e_{HC} and a current I_c , i.e. a threshold magnetic length L_{HC} given through :

$$L_{HC} = (b + e_{HC})\left(1 - \left(\frac{e_c}{e_{HC}}\right)^3\right) \tag{4}$$

$$-\frac{S}{\gamma}b + e_{HC} + \frac{e_c^3}{2e_{HC}^2} - L_{HC}\,\mathrm{Ln}\left(1 + \frac{e_{HC}}{b}\right) = 0 \tag{5}$$

Solving this set of equations is straightforward [14], but the discussion of the extreme cases is physically more illuminating.

a) microscopic regime ($e << b$; $S \approx S_c$)

The spreading parameter $S > 0$ is not large enough to achieve complete wetting if $S < S_c$ but magnetic forces induce the magnetic transition. One finds ($e \ll b$) a film thickness

$$e_{HC} = e_c \sqrt{\frac{S_c}{S}}$$

at a threshold

$$L_{HC} = b\left(1 - \left(\frac{S}{S_c}\right)^{3/2}\right) .$$

b) macroscopic regime

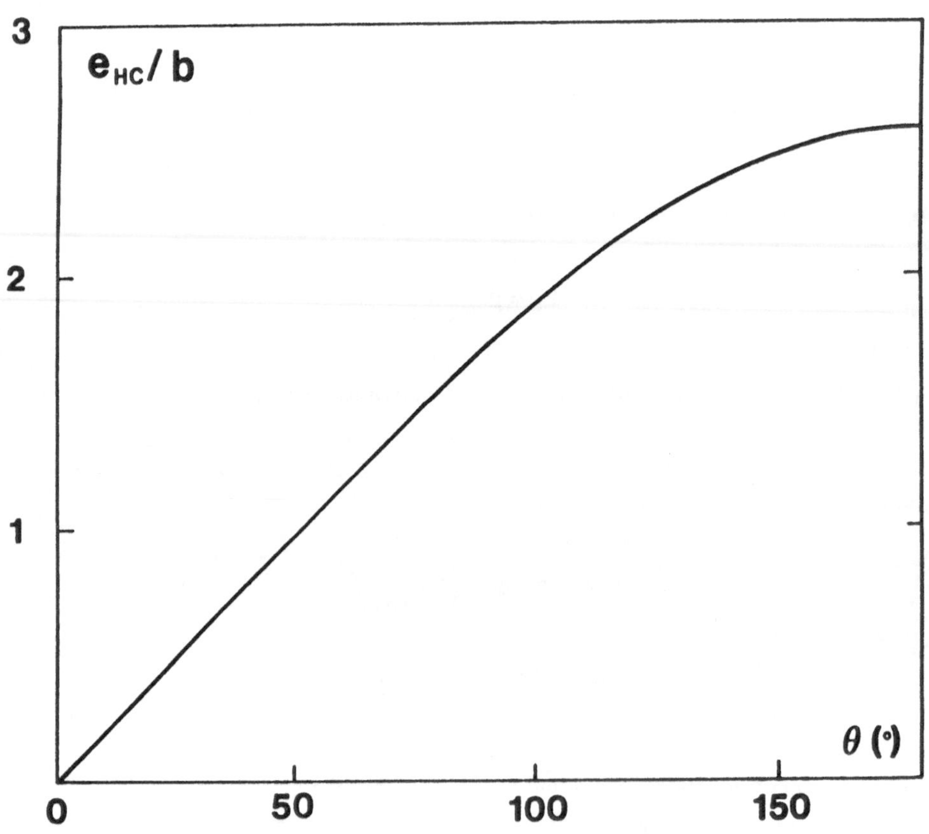

Figure 2 : Non wetting magnetic liquid ($S < 0$ and $\theta \neq 0$) : variations of the reduced magnetic film thickness e_{HC}/b, at the magnetic wetting transition, versus θ, the macroscopic contact angle.

In the macroscopic regime van der Waals forces are negligible : (4) leads to $L_{HC} = b + e_{HC}$ and (5) has only solutions for $S \leq 0$. The film thickness e_{HC} and the threshold L_{HC} of the magnetic wetting transition depends only on the contact angle θ :

$$e_{HC} = L_{HC} - b \tag{6}$$

$$L_{HC} \, \mathrm{Ln}\left(\frac{L_{HC}}{b}\right) - L_{HC} + b \cos\theta = 0 \tag{7}$$

The dependence of e_{HC}/b on θ is given in Figure 2. In this regime there is a balance between magnetic forces and "non-wetting" forces ($S < 0$). As S tends to zero

$$-\frac{S}{\gamma} = 1 - \cos\theta \approx \theta^2/2$$

the film thickness also tends to zero as $e_{HC} \approx b\theta/\sqrt{2}$ and L_H to b. Indeed as e_{HC} tends to zero, we go out of the macroscopic regime and van der Waals forces have to be taken into account.

c) crossover between microscopic and macroscopic regime

Figure 3 is a semi logarithmic plot of the film thickness e_{HC}/b and of the threshold of the magnetic transition L_{HC}/b versus the spreading parameter S within the region where van der Waals forces have to be taken into account (typically $|S| \lesssim 10^{-3}\, S_c$ or $\theta < 2°$ for $S > 0$) ; one can notice the rapid change between macroscopic and microscopic regime in the vicinity of $S = 0$. For $S = 0$, the balance between van der Waals and magnetic forces leads to :

$$e_{HC}^4 = 3\, b e_C^3 \tag{8}$$

and a threshold :

$$L_{HC} = b\left[1 + 2\left(\frac{e_c}{b}\right)^{3/4}\right] \tag{9}$$

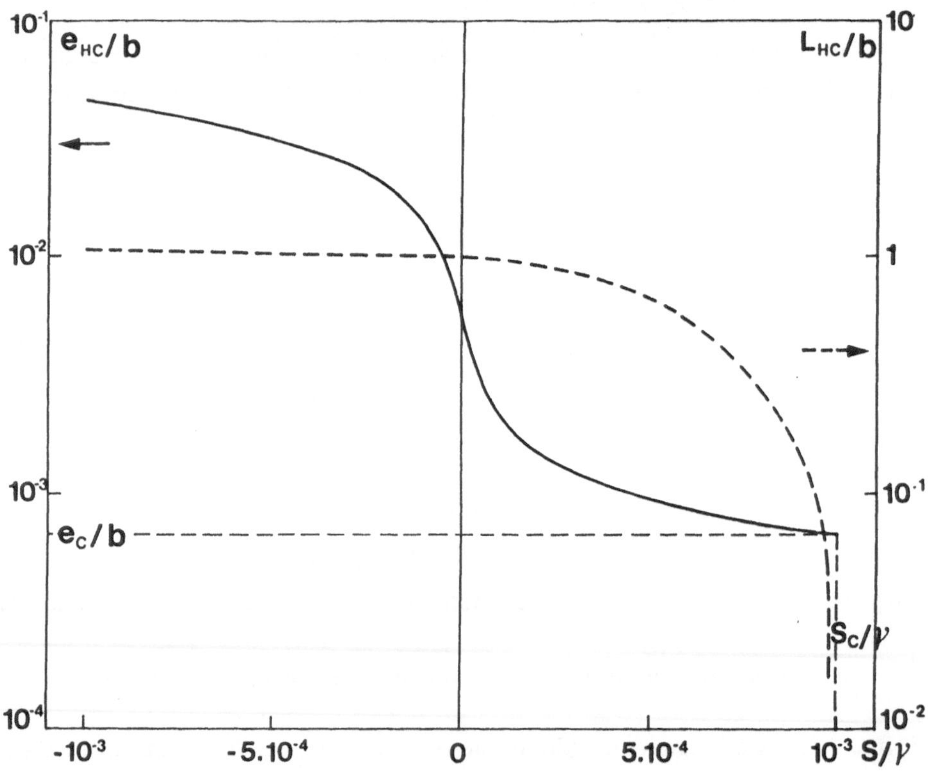

Figure 3 : Semi-logarithmic plot of the wetting film thickness e_{HC}/b and critical magnetic length L_{HC}/b of the magnetic wetting transition versus the reduced spreading parameter S/γ. $S/\gamma = -10^{-3}$ corresponds to a non wetting magnetic fluid with a static contact angle $\theta \approx 2°$. $S/\gamma = 10^{-3}$ corresponds to spontaneous microscopic wetting without magnetic forces ($S = S_c$).

d) experiment

As magnetic fluid, we have used a ferrofluid. Ferrofluids are colloidal dispersions of magnetic particles of typical size 10 nm [10,15]. The ionic ferrofluid [16] that we have chosen is an aqueous solution of Fe_2CoO_4 macroanions which exhibits a large magnetic permeability ($\mu_r = 1 + \chi = 7$). It leads to large magnetic effects even at low fields, i.e. low current in the wire and thus small thermal effects. To prevent evaporation and minimize gravity, the whole system (wire and ferrofluid) has been immersed in freon ; the density difference between the two liquids is $\Delta\rho = 170$ kg m^{-3}. The capillary length directly measured using the Porter method [12] is $K^{-1} = \sqrt{\gamma / \rho g} = 700$ μm leading to $\gamma = 0.8$ mJ m^{-2}. In order to get a large enough magnetic field and a wire radius b much smaller than K^{-1}, we took a 50 μm-radius wire. For I = 1 A, the magnetic field at the wire surface ($B = \mu_0 / 2\pi b$) is 4.10^{-3} T and corresponds to a magnetic length L_{HC} = 120 μm = 2.4 b. Microscope observation at different magnifying factors are made through a video tape recording system ; our typical size accuracy is within a few microns. Height measurements Z are given from a base line we have arbitrarily set at r = 25 b. Pictures (a,b,c,d) correspond to equilibrium shapes obtained at different currents :

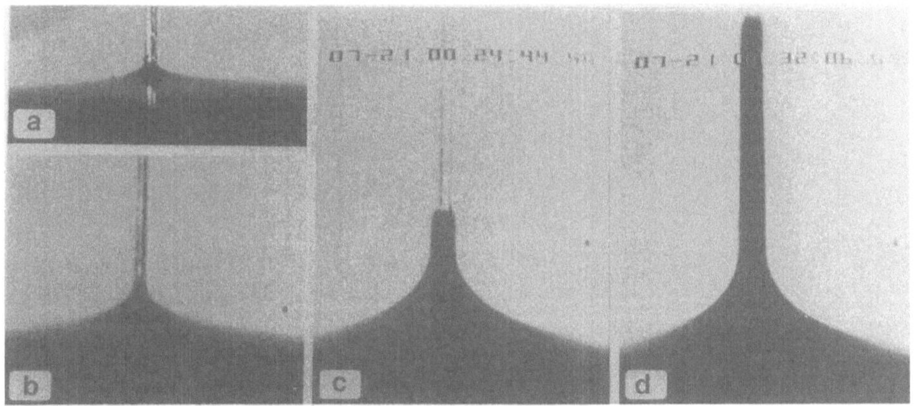

Pictures. - Pictures of the ascension of a ferrofluid, along a vertical wire of radius b = 50 μm. The current I in the wire determines the magnetic length $L_H = \mu_0 \chi I^2/8 \pi^2 \gamma$.

a) I = 0, no field (L_{HC} = 0), the ascension height is of the order of b (Z = 3.5 b) ; the measured contact angle is θ = 20°.

b) I = 600 mA (L_{HC} = 0.9 b) ; Z is enhanced (Z ≈ 10 b) but not drastically.

c) I = 1.020 A (L_{HC} = 2.5 b) ; Z is again increased (Z ≈ 20 b). The apparent contact angle is close to 90° and different from the contact angle at I = 0.

d) I = 1.050 A (L_{HC} = 2.65 b). Just a 6% field enhancement produces such a jump of Z (Z = 55 b !). Compared to picture c, the tip and the lower part of the profile are unchanged but a 40 b long sheath is grown in between, the radius of which decreases smoothly from 2.2 b at the top down to 2.5 b at the bottom.

Clearly, we observe the predicted magnetic wetting transition [17,18,19,]. Moreover, exact calculations [17] of the magnetic fluid profiles (figure 4) are in reasonnable agreement with the observed ones (picture) if and only if the θ variations are taken into account. These θ variations are understood as a dynamical spreading angle ; on a plane $\theta = 90°$ is the dynamic angle [13] for a totally wetting fluid (static angle $\theta = 0$ and $S < 0$) and equilibrium is never reached. This 90° angle is in agreement with the spreading velocity and the shape of the profile [20]. Below the transition, very small sweep-up [20] of the current allows conservation of the contact angle (20°) ; in this case, the sheath thickness close to the transition is small in agreement with (7) and figure 2.

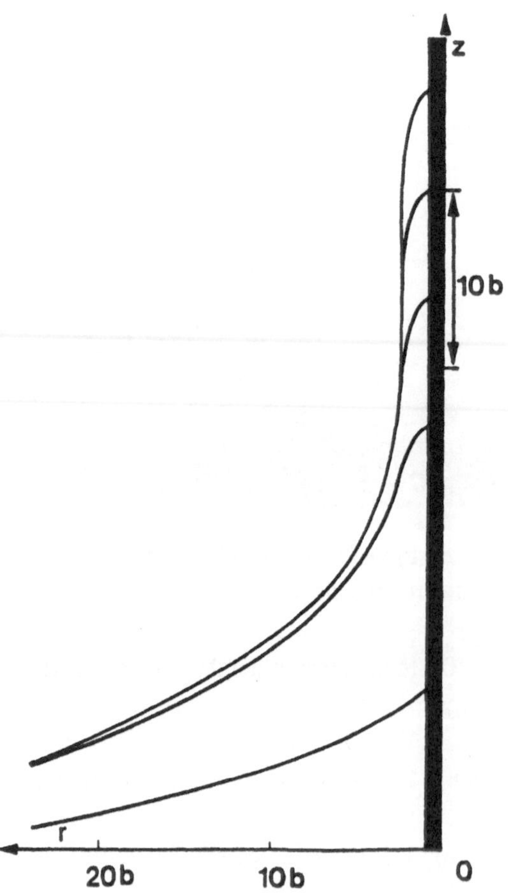

Figure 4 : Computed profiles of the spreading of a magnetic liquid along a conducting wire of thickness b (in black). Different curves correspond to different wire currents and then to different magnetic lengths L_H. From top to bottom $1 - L_H/L_{H_c} = 5.10^{-5}$, 5.10^{-4} , 5.10^{-3} , 5.10^{-2} and 5.10^{-1} . L_{H_c} is the magnetic length threshold of the magnetic wetting transition.

4 Conclusion

We have measured the spreading of a non wetting magnetic fluid along a vertical conducting wire for different wire currents. Our data clearly show that above a given magnetic length threshold, the wire is covered by a sheath of magnetic liquid, the height of which is only limited by gravity. Our data are in reasonable agreement with our prediction : a magnetic wetting transition for a non wetting magnetic fluid analogous to the wetting transition on a thin cylinder, predicted for a completely wetting fluid in the presence of van der Waals forces, but at a macroscopic length scale.

References

1. Young T., Philos. Trans. R. Soc. London **95** (1805) 65.
2. Zisman W., in "Contact Angle, Wettability and Adhesion" edited by F.M. Fowkes, Advances in Chemistry Series N° 43 (American Society, Washington D.C., 1964), p. 1.
3. Cahn J.W., J. Chem. Phys. **66** (1977) 3667.
4. Huh C. and Scriven L.E., J. Colloid and Interface Sc. **35** (1971) 85.
5. De Gennes P.G., Rev. Mod. Physics **57** (1985) 827 and references therein, and Special Issue of "Revue de Physique Appliquée" Vol. 23 (Paris 1988).
6. Cazabat A.M., Contemp Phys., **28**, 347 (1987).
7. Hardy W., Philos. Mag. **38** (1919) 49.
8. Beysens D., Estève D., Phys. Rev. Lett., **54**, 2123 (1985).
9. Pincus P., in "Biophysical effects of steady magnetic fields", (Springer Verlag, 1986).
10. Rosensweig R.E., "Magnetic fluids", Int. Sci. Tech. (1966) 48 ; "Ferrohydrodynamics" (Cambridge University Press, 1985).
11. Brochard F., J. Chem. Phys. **84** (1986) 4664.
12. Adamson A., "Physical Chemistry of Surfaces", John Wiley and Sons Ed., New York, (1982).
13. Joanny J.F., Thèse de Doctorat d'Etat, Université Pierre et Marie Curie, Paris (1985). Joanny J.F. and de Gennes P.G., C.R. Acad. Sc. Paris **299 II** (1984) 279 ; **299 II** (1984) 605.
14. Bacri J.C., Perzynski R., Salin D., C.R. Acad. Sc., Paris **307 II**, 467 (1988).
15. Bacri J.C., Perzynski R. and Salin D., Andeavour, **12**, 76 (1988).
16. Massart R., IEEE Trans. on Magnetics **17** (1981) 1247.
17. Bacri J.C., Perzynski R., Salin D., Tourinho F., Europhysics Lett., **5**, 547 (1988).
18. Bacri J.C., Frenois C., Perzynski R., Salin D., Rev. Phys. Appl., **23**, 1017 (1988).

19. Bacri J.C., Brochard-Wyart F., Di Meglio J.M., Perzynski R., Quéré D. and Salin D., "Hydrodynamics of dispersed media" éd. by J.P. Hulin, A.M. Cazabat, F. Carmona and E. Guyon (to be published).
20. Bacri J.C., Esnault C., Perzynski R. and Salin D., (to be published).

RECENT EXACT RESULTS ON WETTING

D.B. Abraham[1], C.M. Newman[2]

[1] Department of Theoretical Chemistry, 5 South Parks Road
Oxford, OX1 3UB, U.K.
[2] Courant Institute of Mathematical Sciences, Washington Square,
New York, NY 1003, U.S.A

1 Introduction

Broadly speaking, theoretical results in the statistical mechanics of condensed matter derive from the following sources : approximations (such as the renormalization group and mean field theory), exact solutions and rigorous results. The well-known juxtaposition of both physical and mathematical subtlety in this field makes the failure of approximations difficult to interpret unambiguously. This is where exact solutions come in. One constructs a model of the system in question, often in an apparently accidental way, which is amenable to precise mathematical treatment, so that only the implied physics can be inadequate. An exemple of this is Onsager's exact solution of the planar Ising model [19] which indicated that the van der Waals, or classical, theory of phase transitions and critical phenomena is not always even qualitatively correct, since it gives different critical indices. As will be seen later, the status of the van der Waals - Maxwell theory of interface structure has also been questioned by extensions of Onsager's work [2].

The next step is to embed, if possible, the exact solution in question in a class of models with some qualitative similarities. A typical example of this is the description of the low-temperature phases of the planar Ising model in terms of solid – on – solid strings and bubbles, to which we shall return in the next section : such a class of models ought to contain, for instance, a wider range of uniaxial, ferromagnets and their analogues. A further use of this embedding process is to create a language which experimentalists will find useful and to their taste, being more digestible than the exact solutions themselves.

More often than not, even grossly simple models turn out not to be solvable exactly with available techniques. Then one can resort to rigorous perturbation expansions which are only proved to converge for restricted ranges of parameters and thus have less impact than exact solutions, although they usually are more general in character. An example of this is the three-dimensional SOS model of wetting, discussed later in this volume by Bricmont. In the third section, another wetting model will be described which is exactly-solvable and which leads to different results from those of ref. 11.

2 Planar results

Consider a binary mixture which has separated into two coexistent, immiscible phases labelled A and B. The properties of the free interface are well-understood, at least for $d = 2$. This began with the work of Gallavotti [3] who considered the planar Ising model with the Dobrushin [4] boundary conditions applied to a strip of finite width. The long contour connecting the spin reversals on opposite sides of the system at $(-L,0)$ and $(L,0)$ undergoes fluctuations over a distance $\sim\sqrt{L}$, at least for $0 < T < 2^{-32}\, T_c$ (2).

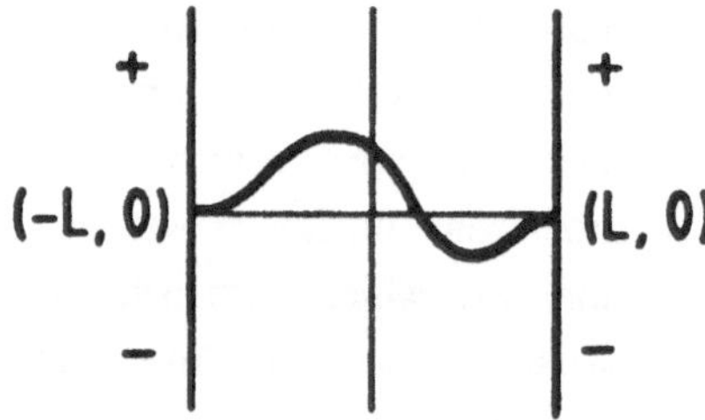

Fig.1 : Planar Ising model with domain wall or interface induced by the Dobrushin boundary conditions

Shortly afterwards, the Dobrushin boundary conditions were analysed exactly for all temperatures. We mention a few of the results [5,6,7] :

Result 1 : The magnetisation in the strip denoted $m(x,y|L)$ satisfies $m(x,y|\infty) = 0\ \forall(x,y)$ and

$$\lim_{L\to\infty} m(\beta L, \alpha L^{\delta}) = \begin{cases} 0 & 0 \le \delta < 1/2 \\ m^{*} \operatorname{sgn} \alpha & \delta > 1/2 \\ m^{*} \operatorname{sgn} \alpha \, \Phi\left(\dfrac{b\,|\alpha|}{\sqrt{1-\beta^{2}}}\right) & \delta = 1/2 \end{cases} \tag{1}$$

for $\beta \in -1,1$, where m^{*} is the spontaneous magnetisation and

$$\Phi(x) = \frac{2}{\sqrt{\pi}} \int_{0}^{x} e^{-u^{2}}\, du \tag{2}$$

with

$$b = \left(\sinh 2\,(K-K^{*})\right)^{1/2}\ ,\ e^{-2K^{*}} = \tanh K \tag{3}$$

Remark : These formulae arise naturally in a coarse-grained theory the exact solution of which justifies approximations of the solid-on-solid type [8,9]. Provided the anisotropy of the surface tension is taken into account, simple fluctuation theory also gives (1) to (3). A full discussion of equilibrium states with general boundary conditions is contained in ref. 2, together with references.

Result 2 : Consider an A – B phase-separating interface which begins and ends in a wall ; this is taken to be the line $y = 0$.
For $\beta \in (-1,1)$, $\delta > 0$ and $\alpha > 0$

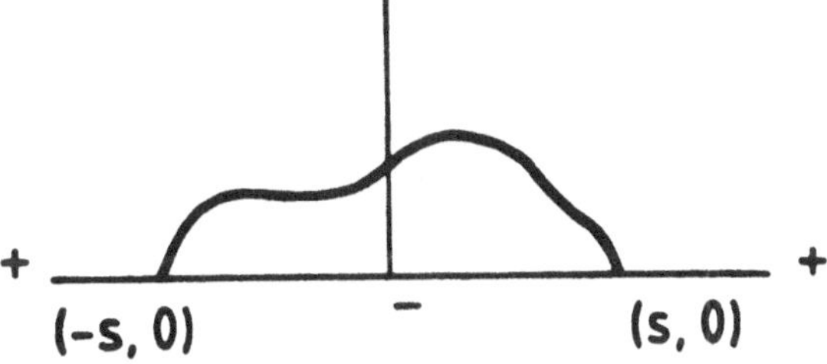

Fig. 2 : Interface near wall.

$$\lim_{s\to\infty} \langle\sigma(\beta s,\alpha s^{\delta})\rangle = \begin{cases} -m^* & 0<\delta<1/2 \\ m^* & \delta>1/2 \\ m^*\,\Psi\left(\dfrac{\alpha b}{\sqrt{1-\beta^2}}\right) & \delta=1/2 \end{cases} \tag{4}$$

where

$$\Psi(x) = 1 - \frac{4}{\sqrt{\pi}}\left\{x\, e^{-x^2} + \int_x^{\infty} e^{-u^2}\, du\right\} \tag{5}$$

Remarks : The incremental free energy is the same whether or not there is the restriction of the interface to the upper half plane. The notion that this is so because the interface wanders away from the wall so far that it does not notice its existence is not born out by equations (1) to (5). In the restricted case the interface clearly makes returns to the wall and is repelled by it. This phenomenon has been termed entropic repulsion [10] and the returns to the surface are called fingers. They play a crucial role in certain phenomenological theories of the wetting transition [1] and in rigorous results [11]. The key idea is that interfaces, even though they are fluctuating strongly interact with walls and with each other. This will form an essential ingredient of the $d = 3$ wetting model described in the next section.

This section concludes with a third planar-Ising result.

Result 3 : Suppose in the system described under result 2 that the vertical bonds at the surface are weakened. On energetic grounds the interface will stick to the wall, but it will loose entropy thereby. A new phase transition emerges at a temperature T_W satisfying $0 < T_W < T_c(2)$. If A interacts more strongly with the wall, for $0 < T < T_W$ we have a microscopic film of B at the wall which becomes macroscopic in thickness on going to $T \geq T_W$. The free energy [7] and length scales have been investigated in detail [7,12], giving a complete picture of a wetting transition in an exact solution.

3 Definition of 3-d wetting model

The phase-separating surface lying above a substrate plane can be modelled by defining a height variable $h(r)$ for each $r \in Z^2$; thus there is an underlying quadratic lattice, the discreteness of which is supposed to represent the molecular character of matter. A number of restrictions on $h(r)$ will be made which allow an exact analysis of the problem ; these are as follows [13,14].

1. $h(r) = 0$ for every r outside a finite $\Lambda \subset Z^2$.

2. $h(r) \in Z$, $h(r) \geq 0$.

3. $h(r) - h(s) = 0, \pm 1$ for all $|r - s| = 1$.

4. Height is conserved round closed loops of Λ.

5. For every $r \in \Lambda$, there is at least one nearest neighbour path joining it to Z^2 / Λ on which the height does not increase.

The final restriction excludes hidden valleys in the surface ; the significance of this will be discussed later.

First we note that there is an alternative description of this model [13] which develops the point of view of Kossel [15] and Stransky [16] in which the phase-separating surface is the top of a pile of units cubes, each representing a molecule. This surface has two important morphological features : the connected sets of points $r \in \Lambda$ which have identical h(r) values are called *terraces*. The lines on the dual lattice Z^{2^*} on which the height changes are called *ledges*. Height conservation demands that they must be closed.

It is straightforward to see that the configurations of ledges compatible with the above rules, when projected down into the substrate plane, are in 1 : 1 correspondence with the Peierls contours of the planar Ising model with plus boundaries. Indeed, this is the aim of the restrictions implied by the model.

We now consider the configurational energy of the surface. Let A be the number of plaquettes (elementary squares) of the surface which are parallel to the substrate. From the construction this is also the area of contact of the bases of the droplets with the under-lying surface. The configurational energy is taken to be

$$E = \tau \sum_{|r-s|=1} | h(r) - h(s) | + (\tau - \varepsilon_1) A_1 \qquad (6)$$

where τ is the surface tension which multiplies the total surface area of the droplets, but ε_1 is the binding to the substrate which is taken here to be a contact interaction.

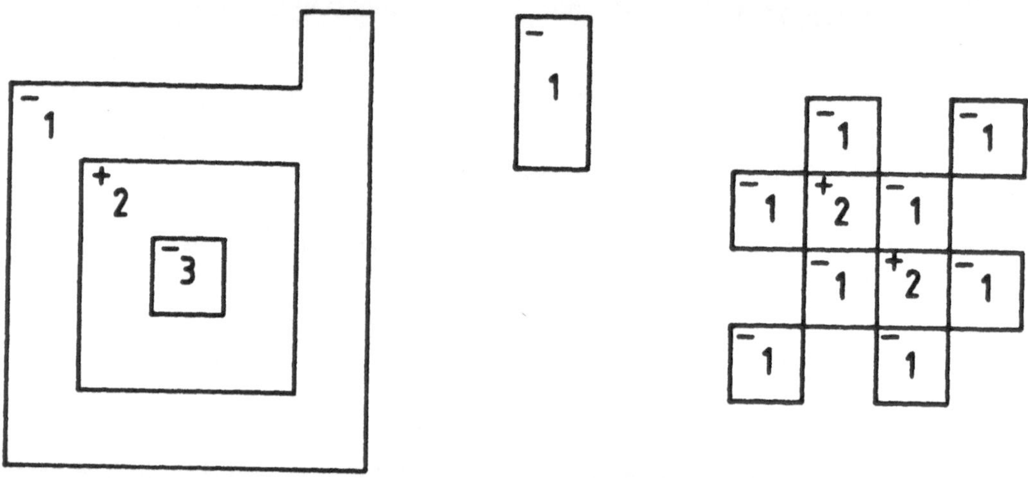

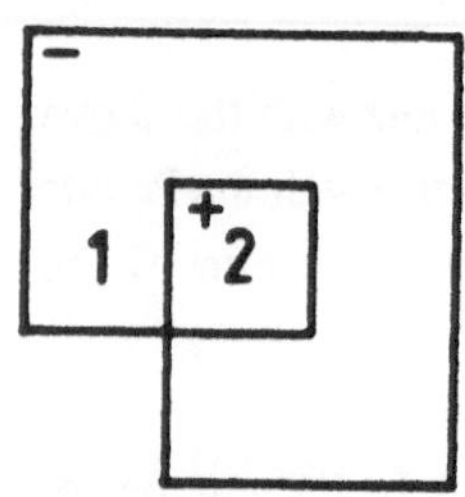

Fig. 3 : Example of allowed configurations, associated Peierls contours and spin configurations.

Using the analogy with spins $\sigma(r) = \pm 1 \ \forall \ r \in \Lambda$, $\sigma(r) = 1 \ \forall \ r \in Z^2/\Lambda$, (6) can be written

$$E = -\frac{\tau}{2} \sum_{|r-s|=1} \left(\sigma(r)\,\sigma(s) - 1\right) - \ b \sum_{r} \left(\mu(r) - 1\right) \tag{7}$$

where the sums are over Λ and $b = \varepsilon_1 - \tau$. The variable $\mu_\Lambda(r) = 1$ (otherwise 0) if some path goes from r to Z^2/Λ with no spin changes. It is thus the indicator for the plus cluster of the boundary in percolation language. When b = 0, we have the planar Ising model with no external field the free energy of which was obtained by Onsager [17]. There is a phase transition at $T = T_c(2)$, where $\sinh[(\tau/T_c(2))] = 1$, with an associated logarithmic divergence of the specific heat.

The interpretation of the phase transition follows by recalling that the terraces are parallel spin clusters, the subject of Ising percolation theory. Moreover h(r) is the minimum number of Peierls contours crossed by any path on Z^2 connecting r to the boundary of Λ ; it is thus related to the size and shape of parallel spin clusters. The key theorem on Ising percolation was proved by Coniglio et al [18].

For $T \geq T_c(2)$, with probability one any parallel spin cluster is finite.

For $T < T_c(2)$, and plus boundary conditions there is a unique infinite plus cluster ; all other clusters are finite.

The first part enables us to prove that, for all $T \geq T_c(2)$, $\lim \Pr_\Lambda h(0) \geq k = 1$ for all $k \geq 0$ and $\lim \langle h(0)_\Lambda \rangle = \infty$. Thus the interface has departed from the substrate, which is covered by a macroscopic film of phase B.

For $T < T_c(2)$, we show that

$$\Pr_\Lambda h(0) \leq k \leq (1 - m^*)^k \tag{8}$$

for all $k \geq 0$, uniformly in Λ, where m^* is the spontaneous magnetisation of the planar Ising model [19]. It follows that

$$\lim \langle h(0) \rangle_\Lambda < (1 - m^*)/m^*.$$

Thus the interface stays near the wall as $\Lambda \to \infty$ and the film of B at the substrate remains microscopic in thickness. The second part of the percolation theorem implies that the droplets in this case of partial wetting are microscopic in size, unlike the usual picture of a macroscopic, sessile drop with a contact angle satisfying the Young rule.

The proof of (8) uses a markovian idea. If $h(0) \geq 1$, then it is surrounded by an outermost Peierls contour γ. Thus for $h \geq 1$

$$\Pr_\Lambda \{h(0) \geq k\} = \sum_\gamma \Pr_\Lambda \{0 \in \text{int}\, \gamma\} \Pr_{\text{int}\, \gamma} \{h(0) \geq k - 1\} \tag{9}$$

We then use a monotonicity property : the left hand side increases in Λ for any k. Apply this on the right and use the result [18] that

$$\Pr_{\Lambda}\{0 \in \text{int}\,\gamma\} \leq (1 - m^*) \qquad (10)$$

Further details of these proofs are given in ref. 14. In addition, even though $b > 0$ is special type of external magnetic field, the problem can still be solved [13]. The phase transition is of Onsager type and occurs at $T = T_c(2)$ independently of b. The region $T \geq T_c(2)$, $b \geq 0$ is always wet, whereas $T < T_c(2)$, all b is always partially wet. As yet, the behaviour for $T \geq T_c(2)$, $b < 0$ is unknown. It may be there is a critical $b = b_c$ such that for all $T > 0$, $b < b_c$ the interface is far form the wall.

$[0, T_c(2)]$ is complete, but we believe that the behaviour in $(T_c(2), \infty)$ is of more mathematical interest. The singularity of the free energy as $T \to T_c(2)+$ is probably of the de-roughening type.

Remarks

1. The standard model for wetting is the SOS one with $h(r) \in Z$, $h(r) \geq 0$ and energy

$$E = \tau \sum_{|r-s|=1} |h(r) - h(s)|^p - \sum \delta_{h(r),0} \qquad (11)$$

The phase transition occurs because of a detailed interplay between the entropic repulsion of the interface from the wall and its attraction thereto caused by the second term in (11). Because of property 5 in the definition of our model, such an effect does not occur ; rather, there is an entropic repulsion between the ledges, although this does not play an overt role in the analysis. It is possible that for $b < 0$ the models will appear much more alike and even give the same behaviour.

2. The complete exclusion of hidden valleys in our model is clearly an undesirable restriction. Equally well, so is the perfect up-down symmetry in the standard SOS model in the partially-wet phase. An advantage of this symmetry in that the SOS model can describe a free interface between phases [20] whereas ours does not.

3. It can be shown that [14]

$$\lim_{T \to T_c(2)-} \lim_{\Lambda \to \infty} \langle h(0,0) \rangle \uparrow \infty .$$

Thus the wetting-transition description in the low-temperature phase is complete.

ACKNOWLEDGEMENTS

Some of the planar Ising results described above were conjoint work of one of the authors with Issigoni and Reed. We should also like to acknowledge M.E. Fisher, M.P. Gelfand, Ch. Ed. Pfister, V. Privman and N.M. Svrakic. We also thank the organisers of this workshop, Ph. de Gottal, F. Dunlop and J. De Coninck, for the invitation to present these results.

REFERENCES

1. S. Dietrich in Phase Transitions and Critical Phenomena ed. C. Domb and J.L. Lebowitz (Academic Press) vol. 12 (1988).
2. D.B. Abraham in Phase Transitions and Critical Phenomena, ed. C. Domb and J.L. Lebowitz (Academic Press) vol. 10 (1986).
3. G. Gallavotti, Commun. Math. Phys. **27**, 103 (1972).
4. R.L. Dobrushin, Theory Prob. Appl., **17**, 582 (1972) ; **18**, 253 (1973).
5. D.B. Abraham and P. Reed, Phys. Rev. Letts., **33**, 377 (1974) ; Commun. Math. Phys., **49**, 35 (1976).
6. D.B. Abraham and M.E. Issigoni, J. Phys., **A13**, L89 (1980).
7. D.B. Abraham, Phys. Rev. Letts, **44**, 1165 (1980).
8. D.S. Fisher, M.P.A. Fisher and J.D. Weeks, Phys. Rev. Lett., **48**, 369 (1982).
9. D.B. Abraham, Phys. Rev. Letts., **47**, 545 (1981).
10. M.E. Fisher and D.S. Fisher, Phys. Rev., **25**, 3192 (1982).
11. J. Bricmont, A. El Mellouki and J.M. Fröhlich, J. Stat. Phys., **42**, 743 (1986).
12. D.B. Abraham and L.F. Ko, Phys. Rev. Lett. **63**, 275 (1989).
13. D.B. Abraham and C.M. Newman, Phys. Rev. Letts., **61**, 1969 (1988).
14. D.B. Abraham and C.M. Newman, to appear in Commun. Math. Phys., Dobrushin Festschrift number, **125**, 181 (1989).
15. W. Kossel, Nachr. Ges. Wiss. Göttingen, Mathemat./Physikal. Klasse S. 135 (1927).
16. J.N. Stransky, Z. Phys. Chemie., **136**, 259 (1928).
17. L. Onsager, Phys. Rev., **65**, 117 (1944).
18. A. Coniglio, C.R. Nappi, F. Peruggi and L. Russo, Commun. Math. Phys., **51**, 315 (1976).
19. C.N. Yang, Phys. Rev., **85**, 808 (1952).
20. J.M. Fröhlich and T. Spencer, Commun. Math. Phys., **81**, 527 (1981).

WETTING OF A DISORDERED SUBSTRATE

H. Orland

Service de Physique Théorique, Laboratoire de l'Institut de Recherche Fondamentale du Commissariat à l'Energie Atomique de Saclay, F-91191 Gif-sur-Yvette Cedex

The purpose of this talk is to discuss the effect of randomness on the wetting transition. More precisely, we shall only discuss the effect of random variations (at the microscopic level) in the strength of the substrate potential on the transition, leaving to Th. M. Nieuwenhuizen the discussion of bulk randomness[1]. This talk is a brief review of a serie of works published in references (2,3).

1 Critical wetting

The effect of randomness on a critical point can be discussed qualitatively using the Harris criterion[4]. Assuming that h_i is a scaling field, with crossover exponent ϕ, this criterion states that randomness in h_i is relevant, marginal or irrelevant if the combination of exponents $\nu d - 2\phi$ is respectively negative, zero or positive. This can be easily seen by noting that near a critical point, the singular part of the free energy takes the scaling form:

$$f \sim |t|^{2-\alpha} \varphi\left(\frac{h_i}{|t|^{\phi}}\right) \tag{1}$$

where t is the reduced temperature.

The quenched free energy can be expanded for small disorder as:

$$f \sim f_0 + \frac{1}{2}\Delta^2 |t|^{2-\alpha-2\phi} \varphi''(0) \tag{2}$$

where the moments of h_i satisfy:

$$\begin{aligned} \bar{h}_i &= 0 \\ \overline{h_i h_j} &= \delta_{ij}\Delta^2 \end{aligned}$$

The Harris criterion follows trivially from (2).

At critical wetting, there are two relevant fields[5] :

i) the bulk magnetic field

ii) a temperature-like field, which is a combination of the actual temperature and of the surface magnetic field.

Thus, randomness in the surface field amounts to randomness in the temperature-like field. The associated crossover exponent is $\phi = 1$, and from the hyperscaling relation, the relevance of randomness is related to the sign of the exponent α : negative α implies irrelevant, whereas $\alpha > 0$ implies relevant.

We distinguish two cases:

i) Short range forces. The upper critical dimension (i.e. dimension above which mean-field is valid) is $d^* = 3$.

a) $d > 3$. Mean-Field yields $\alpha = 0$. Since the specific heat has a finite jump, disorder is *marginally irrelevant*.

b) $d = 3$. It was shown in Ref.(6) that $\alpha < 0$. Thus, disorder is *irrelevant*.

c) $d = 2$. From D.Abraham's[7] results, it is known that again $\alpha = 0$, with a finite discontinuity in the specific heat. Thus, disorder is marginally irrelevant.

ii) Long range forces. The upper critcal dimension is $d^* = \frac{11}{5}$. Thus, for $d > \frac{11}{5}$, (the physically interesting case) mean field gives $\alpha = -1$, and disorder is irrelevant.

Therefore, to be complete, we must study in more detail the case of short range forces, both in MF and in $d = 2$.

2 First order wetting

In the case of first order wetting, there is a prewetting line, off-coexistence, terminated by a critical prewetting point (Figure 1).

In bulk dimension d, the prewetting line is a first order line, in the universality class of the Ising model in dimension $(d-1)$. Indeed, the width w of the wetting layer is a diatonic variable (thin-thick), defined on a $(d-1)$ dimensional space (interface). Let us discuss specifically the case of $d = 3$ in the bulk. The effect of a surface random field on the critical prewetting point T_c is in the same universality class as the random field Ising model in $d = 2$ (RFIM). It seems that $d = 2$ is the marginal dimension of the RFIM[8]. However, it is not known precisely whether long range order remains, or is destroyed in $d = 2$. We thus have two possibilities:

i) If the transition is destroyed, there is no more jump along the prewetting line, but a rapid continuous variation in the width of the wetting layer. The wetting point becomes continuous for infinitesimal disorder.

ii) If the transition remains, it falls into a new universality class, and thus, wetting remains first order, but the critical prewetting point charges its nature.

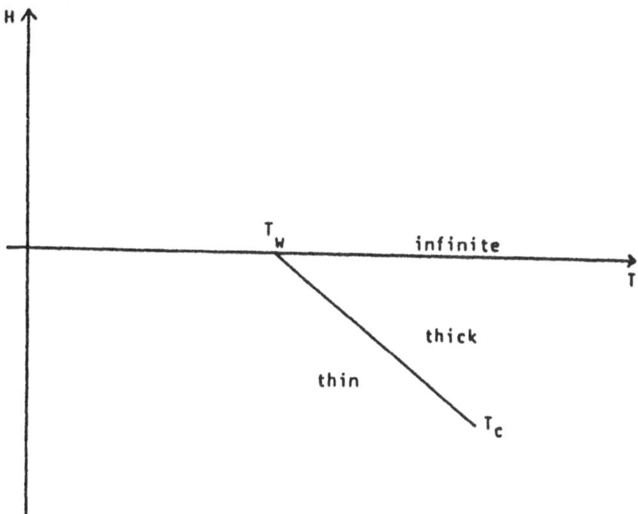

Fig. 1. Phase diagram of the wetting transition for the case of first-order wetting

3 Mean field theory

Within mean-field theory (MFT), the quenched free energy can be computed as follows. We introduce replicas[9], according to the equation:

$$\text{Log } Z = \frac{\mathrm{d}}{\mathrm{d}n}\overline{Z^n}\bigg|_{n=0}$$

Assuming that the surface field h_i are independent random variables, with probability $P(h_i)$, we obtain[2] :

$$\overline{Z^n} = \int \prod_{i=1}^{N}\prod_{\alpha=1}^{n} \mathrm{d}\varphi_i^\alpha \mathrm{e}^{S(\varphi_i^\alpha)} \tag{3a}$$

where

$$\begin{aligned} S(\varphi_i^\alpha) = &-\frac{1}{2}\sum_{i,j}\sum_{\alpha,1}^{n} \varphi_i^\alpha J_{ij}\varphi_j^\alpha \\ &+\sum_i \text{Log}\left\langle \prod_{\alpha=1}^{n} (2\cosh(\sqrt{\beta}\varphi_i^\alpha + \beta h_i))\right\rangle \end{aligned} \tag{3b}$$

and the brackets denote an average over the disorder:

$$\langle A\{h_i\}\rangle = \int \prod_{i=1}^{N} \mathrm{d}h_i P(h_i)\cdot A(h_i)$$

Mean field is obtained by minimizing the action S over the fields $\{\varphi_i^\alpha\}$, assuming that there is no replica symmetry breaking, i.e. $\varphi_i^\alpha = \varphi_i$ independent of α.

Introducing the magnetization m_i at site i, the mean-field equation becomes, for weak disorder $(\beta\Delta_j \ll 1)$

$$\mathrm{th}^{-1} m_i = \beta \sum_j J_{ij} m_j + \beta H_i - \beta^2 \Delta_i m_i \tag{4}$$

A. PREWETTING

It is easily seen that the critical prewetting temperature is given by:

$$T_c(\Delta) = T_c(0)\left[1 - \left(\frac{\Delta}{T_c(0)}\right)^2 + \cdots\right]$$

and so, it is decreased by weak random surface fields.

If the exact result of the random field Ising model be that a transition remains in weak randomness, then this result is at least qualitatively correct, and indicates a lowering of the critical temperature. If the correct result is instead that infinitesimal disorder destroys the transition, then MFT misses this result, but does shift T_c in the correct direction.

B. FIRST-ORDER WETTING

Taking a 3-dimensional semi-infinite system, with transverse direction z, the mean-field equations become, in the continuum limit:

$$\beta J \frac{\mathrm{d}^2 m}{\mathrm{d}z^2} + 6\beta J\ m(z) = \mathrm{th}^{-1} m(z) \tag{5a}$$

which is to be solved subject to the boundary condition:

$$\frac{\mathrm{d}m}{\mathrm{d}z} + \frac{h}{J} - m(0) C\left(\beta\Delta^2\right) = 0 \tag{5b}$$

where

$$C\left(\beta\Delta^2\right) = 1 - 4D + \frac{\beta\Delta^2}{J} \tag{5c}$$

There is a continuous wetting transition at β_w, given by[2] :

$$\mathrm{th}^{-1}\left(\frac{h}{JC\left(\beta_w\Delta^2\right)}\right) = 6\frac{h\,\beta_w}{C\left(\beta_w\Delta^2\right)}$$

provided that,

$$\left[C^2(\beta_w\Delta^2) + 6\right]\beta_w J \geq \left[1 - \left(\frac{h}{JC(\beta_w\Delta^2)}\right)^2\right]^{-1} \tag{6}$$

Otherwise, the transition is first order. According to Eq.(5c), C is increased by randomness. Thus, if the pure system has a first-order wetting transition because $C(0)$ is too small for Eq.(6) to be fulfilled, randomness can drive the transition towards critical wetting. Thus, in MFT calculation, a first order wetting transition is not driven continuously infinitesimal randomness. This is completely consistent with the previous MFT result that the prewetting line is not destroyed by infinitesimal randomness. In both cases, a finite width Δ is required.

4 Two-dimensional systems

There remains the question of wetting of a 2-dimensional with random surface field. This has been studied within the framework of a restricted solid-on-solid model (RSOS). It is indeed known that provided the wetting transition is not too close to the critical point, interface models yield exactly the same physics as Ising models[10].

The partition function for the semi-infinite 2-dimensional system reads:

$$Z(\{u_i\}) = \sum_{\{h_i=\{0,1,\cdots,\infty\}\}} \exp\Big[-\beta J \sum_{i=1}^{N} |h_{i+1} - h_i|^{\infty} + \beta \sum_{i=1}^{N} u_i \delta(h_i)\Big] \tag{8}$$

where u_i is the random Gaussian surface field acting on the wall, at $h_i = 0$, with $\bar{u}_i = u_0$ and $\Delta u_i = v$.

The ∞-power in (8) forces $|h_{i+1} - h_i| = 1, 0$ which is the RSOS constraint.

Introducing replicas and performing the quenched average, one gets:

$$\begin{aligned}\overline{Z^n} = \sum_{\{h_i^\alpha\}} \exp\Big[-\beta J \sum_{\alpha=1}^{n}\sum_{i=1}^{N} |h_{i+1}^\alpha - h_i^\alpha|^{\infty} + \beta\bar{u} \sum_{\alpha=1}^{n}\sum_{i=1}^{N} \delta(h_i^\alpha)\Big] \\ + \beta^2 v \sum_{1\leq\alpha<\beta\leq n} \delta(h_i^\alpha)\delta(h_i^\beta\end{aligned} \tag{9}$$

where $\bar{u} = u_0 + \beta\frac{v}{2}$.

The quenched free energy is given by $-\beta F = \frac{\partial}{\partial n}\overline{Z^n}\big|_{n=0}$. The critical behaviour of this model can be computed exactly, by resumming the dominant divergences of the perturbation expansion of (9) in powers of $g = \beta^2 v$. The salient results are:

i) the critical temperature of the quenched system ($n \longrightarrow 0$) is identical to that of the annealed system ($n = 1$) :

$$e^{\beta_c \bar{u}} = \frac{1+2t}{1+t}$$

where $t = e^{-\beta J}$.

ii) the free energy acquires a singular part (which vanishes at the transition), given by

$$-\beta F_{\text{sing}} = \frac{2\pi\mu^2 t}{1+2t} \cdot \frac{1}{\text{Log}\dfrac{1}{\mu}}$$

where $\mu = \frac{1}{T_c - T}$.

This $\frac{1}{\mathrm{Log}\frac{1}{\mu}}$ singularity also shows up in the specific heat.

iii) the divergence of the width of the wetting layer is unaffected by randomness.

These results have been checked numerically by transfer matrix methods, using matrices with $h_{\max}$ up to 500. The number of matrices entering the product is $N = 10^6$ to 10^7. The numerics confirms our results, although the $\mathrm{Log}\frac{1}{\mu}$ singularity has not been seen, due to the too small sizes of the matrices used.

References

1 Th. M. Nieuwenhuizen, these proceedings.

2 G. Forgacs, H. Orland and M. Schick, Phys. Rev. B32, 4683 (1985).

3 a)G. Forgacs, J. Luck, Th. M. Niewenhuizen and H. Orland, Phys. Rev. Lett. 57, 2184 (1986) b)G. Forgacs, J.M. Luck, Th. M. Nieuwenhuizen and H. Orland, J. Stat. Phys. 51, 29 (1988).

4 A.B. Harris, J. Phys. C17, 1671 (1974).

5 N. Nakanishi and M.E. Fisher, Phys. Rev. Lett. 49, 1565 (1982).

6 E. Brézin, B.I. Halperin and S. Leibler, Phys. Rev. Lett. 50, 1387 (1983); R. Lipowsky, P.M. Kroll and R.K. P. Zia, Phys. Rev. B27, 4499 (1983).

7 D. Abraham, Phys. Rev. Lett. 44, 1165 (1980).

8 J.Z. Imbrie, Phys. Rev. Lett. 53, 1747 (1984) and references therein.

9 S.F. Edwards and P.W. Anderson, J. Phys. F5, 965 (1975).

10 J.M.J. van Leeuwen and H.J. Hilhorst, Physica 107A, 319 (1981); S.T. Chui and J.D. Weeks, Phys. Rev. B23, 2438 (1981).

AN INTRODUCTION TO A MATHEMATICAL DESCRIPTION OF THE WETTING PHENOMENA IN THE ISING MODEL

Charles-Edouard Pfister

Département de Mathématiques
Ecole Polytechnique Fédérale
CH-1015 LAUSANNE, SUISSE

1 Introduction

This lecture is devoted to the presentation of mathematical results concerning the wetting of a wall. It is based on a paper by Abraham [1] and our four papers written in collaboration with J. Fröhlich and O. Penrose [2], [3]. Let us suppose that we have a system which can be in two different phases, e.g. a binary mixture. The physical parameters are chosen in such a way that the system is in the coexistence region. The two different phases are called +phase and - phase. Let us prepare the system in the -phase and let us introduce an horizontal wall in the system which adsorbs preferentially the +phase. In some situation we can observe the formation of a thin film of the +phase between the wall and the -phase so that the -phase cannot reach anymore the wall. This is the phenomenon of complete wetting of the wall. Let us notice that the total amount of the +phase is here not fixed a priori. The appropriate statistical ensemble is a grand canonical ensemble. Let us denote τ^- the contribution of the wall to the surface free energy of the system in presence of the -phase in the bulk. In the case of complete wetting it is reasonable to expect that τ^- can be decomposed into two terms : $\tau^- = \tau^+ + \tau^\pm$. The term τ^+ comes from the fact that the wall is actually in contact with the +phase. The second term is the usual surface tension of an horizontal interface between the + and -phases. When the complete wetting does not occur a stability argument shows that $\tau^- < \tau^+ + \tau^\pm$. Of course we can imagine a situation where the roles of the + and - phases are interchanged. On the basis of this simple argument we expect the fundamental relation

$$|\tau^+ - \tau^-| \leq \tau^\pm \tag{1.1}$$

with equality only in the case of complete wetting.

It is possible to imagine another situation. We again have the system in the -phase in presence of the wall but we now suppose that $\tau^- < \tau^+ + \tau^\pm$. We put a (macroscopic) droplet of the +phase on the wall. Since we do not have complete wetting of the wall by the +phase the droplet does not spread completely. We have the situation of the figure 1 below and the contact angle θ satisfies the inequalities $0 < \theta < \pi/2$. (The case $\pi/2 < \theta < \pi$ corresponds to a wall adsorbing preferentially the -phase).

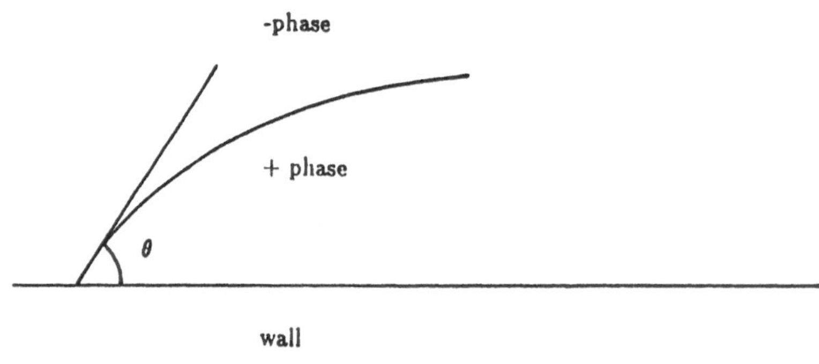

Fig. 1. A droplet of + on the wall.

If the system is isotropic, then the contact angle θ satisfies Young's relation

$$\cos\theta \; \tau^{\pm} = \tau^{-} - \tau^{+}$$

From this relation we get again the fundamental inequality (1.1). A similar conclusion also holds in the anisotropic case like in the Ising model. This situation has been analyzed by De Coninck and Dunlop [4]. (In formula (1.1) $\tau^{\pm}$ refers to the surface tension of an horizontal interface). Here the total amount of the +phase is given a priori. The relevant ensemble is now a canonical ensemble.

The situation above can be analyzed in a semi-infinite Ising model. Indeed, this model can be interpreted as a binary mixture. The interaction with the wall is described by a boundary magnetic field. Once the microscopic hamiltonian is fixed, all relevant questions can in principle be answered, in the appropriate ensemble, on the basis only of the first principles of statistical mechanics. It is true that the problems which have to been solved are usually very difficult, and one is led in most cases to introduce approximate descriptions, e.g. mean field approximations, or special models of interfaces, e.g. solid-on-solid models. Almost all our knowledge and understanding of the wetting phenomena come from the study of such approximations. It turns out however, if we restrict the discussions to simple models like the Ising model, that we can analyze some aspects of the wetting without such approximations. The results are of two kinds : quantitative results, e.g. the exact computation of a magnetization profile, and qualitative results, e.g. analyticity properties of the surface free energy. Almost all quantitative results have been obtained for dimension two. They have been thoroughly reviewed by Abraham [5]. On the other hand many qualitative results are valid for dimensions greater or equal to two.

The main results of the papers mentionned at the beginning form the bulk of the lecture. They all treat the grand canonical ensemble. In 1988 R. Dobrushin, R. Kotecky and S. Shlosman announced new results on the typical configurations of the canonical ensemble of the two-dimensional Ising model with periodic boundary conditions [6]. Using these results it is possible to understand what happens in a canonical semi-infinite Ising model, and to study the shape of a macroscopic droplet like in the figure 1. The mathematical analysis is currently done in collaboration with R. Kotecky. Such results have been first obtained by De Coninck, Dunlop, Rivasseau [7].

The workshop on wetting phenomena organized in Mons had a special character, since experimental, theoretical and mathematical physicists as well as chemists and engineers were present. We had a unique chance to attend lectures on very different viewpoints of the same subject. In the next pages only basic mathematical results are presented in a nontechnical manner. This text is written for the nonspecialists; it is not a review.

2 D-Dimensional Ising model

The d-dimensional Ising model is defined on the lattice $\mathbb{Z}^d$; in absence of an external magnetic field the Hamiltonian is formally

$$H = -J \sum_{<ij>} \sigma(i)\sigma(j) \tag{2.1}$$

where J is a positive coupling constant (ferromagnetism) and $< ij >$ means that we sum only over the nearest neighbour pairs of points of the lattice. The spin variable $\sigma(i) = \pm 1$. It is well-known that the model has two different phases with spontaneous magnetization at low temperature if $d \geq 2$. The mathematical description of these phases is as follows. The model is first contained in a box $\Lambda(L)$,

$$\Lambda(L) = \{i = (i_1, \cdots, i_d) \in \mathbb{Z}^d : |i_k| \leq L\ ,\ \forall k\}$$

Two probability measures for this finite system are introduced, μ_L^+ and μ_L^-. They differ by the choice of a boundary condition. Let us fix the value of the spins outside $\Lambda(L), \sigma(i) = +1, \forall i \notin \Lambda(L)$. In (2.1) we restrict the sum over all nearest neighbour pairs $< ij >$ which have a nonempty intersection with $\Lambda(L)$. The sum in (2.1) is therefore well-defined and its value is denoted $H_L^+(\sigma)$ where σ is a spin configuration (inside $\Lambda(L)$). The measure μ_L^+ is then simply the usual Gibbs measure constructed with $H_L^+(\sigma)$,

$$\mu_L^+(\sigma) = \frac{exp(-\beta H_L^+(\sigma))}{Z_L^+} \tag{2.2}$$

where β is the inverse temperature and the normalization factor Z_L^+ is the partition function

$$Z_L^+ = \sum_\sigma exp(-\beta H_L^+(\sigma)) \tag{2.3}$$

We can compute the free energy of this system,

$$F_L^+ = -\frac{1}{\beta} ln Z_L^+ \tag{2.4}$$

and any expectation values of local observables, e.g. if $i \in \Lambda(L)$,

$$< \sigma(i) >_L^+ = \sum_\sigma \mu_L^+(\sigma).\sigma(i) \tag{2.5}$$

The measure μ_L^- is defined similarly with a -boundary condition, $\sigma(i) = -1 \forall i \notin \Lambda(L)$. In statistical mechanics the volume of $\Lambda(L)$, which is here by definition the number of sites inside $\Lambda(L)$, $|\Lambda(L)| = (2L+1)^d$, is very large, and it is natural to study the model in the asymptotic situation where $L \to \infty$, the so-called thermodynamic limit. If $\beta < \beta_c(d)$, the inverse critical temperature, the two probability measures μ_L^+ and μ_L^- have the same limit as $L \to \infty$. On the other hand, if $\beta > \beta_c(d)$, i.e. at low temperature, we get two different limits, $\mu^+ \neq \mu^-$. In particular,

$$lim_{L\to\infty} < \sigma(i) >_L^+ = M^*(\beta) \tag{2.6}$$

and

$$lim_{L\to\infty} < \sigma(i) >_L^- = -M^*(\beta) \tag{2.7}$$

where M^* is the spontaneous magnetization of the model. The phase transition appears as an instability with respect to the choice of the boundary condition (see [8]) for more details). We see that the +boundary condition is associated with the +phase and the -boundary condition with the -phase. Finally let us introduce the bulk free energy per unit volume,

$$f(\beta) = lim_{L\to\infty} \frac{1}{(2L+1)^d} F_L^+(\beta) \tag{2.8}$$

We can write

$$\begin{aligned} Z_L^+ &= exp(-\beta F_L^+) \\ &= exp(-\beta(2L+1)^d f - \beta O(L^{d-1})) \end{aligned} \tag{2.9}$$

where $O(L^{d-1})$ contains the boundary contribution to the free energy F_L^+. The same formula (2.9) holds for F_L^- since by symmetry $Z_L^+ = Z_L^-$.

Besides the +boundary condition and the -boundary condition which lead to the mathematical definition of the pure phases of the model, there is another boundary condition, the $\pm$ boundary condition, which leads to the definition of the surface tension of an horizontal interface. It is defined as follows : the values of the spins $\sigma(i)$ with $i \notin \Lambda(L)$ is (see figure 2)

$$\begin{aligned} \sigma(i) &= +1 \ if \ i \notin \Lambda_L \ and \ i_d \geq 0 \\ \sigma(i) &= -+1 \ if \ i \notin \Lambda_L \ and \ i_d < 0 \end{aligned} \tag{2.10}$$

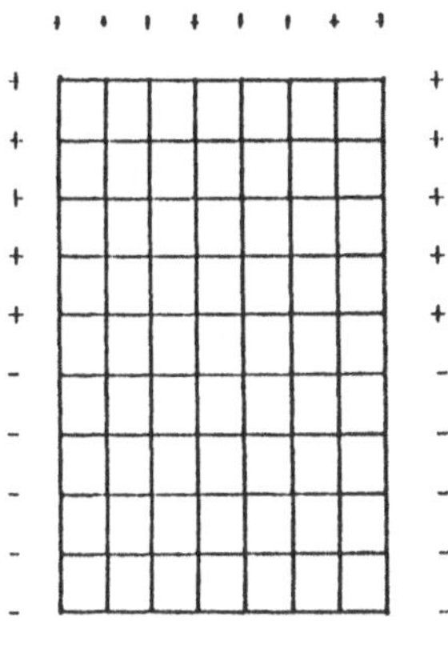

Fig. 2. Boundary conditions

In the coexistence region i.e. for $\beta > \beta_c(d)$ this boundary condition leads to an equilibrium state of the system where the -phase occupies the bottom of the box and the +phase the top of the box. We can use this boundary condition in order to study the properties of an (horizontal) interface between the two pure phases. Let $Z_L^{\pm}$ be the corresponding partition function. Since $H_L^{\pm}$ and H_L^{+} differ only from a boundary term,

$$lim_{L\to\infty} - \frac{1}{\beta}\frac{1}{(2L+1)^d} lnZ_L^{\pm} = f$$

However the contributions of order $O(L^{d-1})$ are in general different. The difference (per unit (d-1)-volume) is given by

$$\tau^{\pm} = lim_{L\to\infty} - \frac{1}{\beta}\frac{1}{(2L+1)^{d-1}} ln\frac{Z_L^{\pm}}{Z_L^{+}} \tag{2.11}$$

The origin of $\tau^{\pm}$ is due to the presence of an interface. This quantity is therefore interpreted as the surface tension of an (horizontal) interface. It is possible to give a more detailed justification for this interpretation. We refer to [8]. We shall see below that $\tau^{\pm}$ has exactly the properties which we expect for the surface tension. In particular, [9]

$$\begin{aligned} \tau^{\pm}(\beta) > 0 \ \ if \ \ \beta > \beta_c(d) \\ \tau^{\pm}(\beta) = 0 \ \ if \ \ \beta \leq \beta_c(d) \end{aligned} \tag{2.12}$$

The last equality is indeed consistent with the fact that for $\beta \leq \beta_c(d)$ there is only one phase and therefore no interface.

3 D-dimensional semi-infinite Ising model. Grand canonical ensemble

In order to simulate the introduction of a wall in the system, we consider the model on a semi-infinite lattice$\mathbb{L}$

$$\mathbb{L} = \mathbb{Z}^{d-1} \times \mathbb{Z}^{+} = \{i \in \mathbb{Z}^{d}; i_d \geq 0\}, d \geq 2$$

and we add to the Hamiltonian (2.1) a boundary magnetic field describing the properties of the wall. The Hamiltonian becomes

$$-J \sum_{<ij>} \sigma(i)\sigma(j) - h \sum_{i:i_d=0} \sigma(i) \tag{3.1}$$

If h is positive the wall adsorbs preferentially the +phase and if h is negative it adsorbs preferentially the -phase.

The first quantity to define is the surface free energy of the model. Since it is a surface free energy we must be careful in the choice of the boundary conditions which we use for the partition function. For example the use of the free boundary conditions is not suitable. Indeed, we would get with such a boundary condition a singularity in h at $h = 0$ for all $\beta \geq \beta_c(d)$. However, for $\beta > \beta_c(d)$ there is no phase transition in the model at $h = 0$. Since there are two pure phases in the model, we must consider two surface free energies : $\tau^+, resp. \tau^-$, the surface free energy when there is in the bulk the +phase, resp. the -phase. They are defined through the choice of the +boundary condition, resp. the -boundary condition. Let us consider the specific case of the -boundary condition. We first constrain, as before, the model in a finite box

$$\Omega(L) = \Lambda(L) \cap \mathbb{L} \tag{3.2}$$

and we impose the values of all spins $\sigma(i)$ with $i \in \mathbb{L}/\Omega(L)$,

$$\sigma(i) = -1 \;\; if \;\; i \in \mathbb{L}/\Omega(L) \tag{3.3}$$

If we restrict the summation in (3.1) over all pairs $< ij >$ such that $< ij > \cap \Omega(L) \neq \phi$ and to all $i \in \Omega(L)$ for the second sum, then we get a well-defined Hamiltonian which we denote $\hat{H}_L^-(\sigma)$ for all configurations σ. The corresponding Gibbs measure is

$$\nu_L^-(\sigma) = \frac{exp(-\beta \hat{H}_L^-(\sigma))}{\hat{Z}_L^-} \tag{3.4}$$

where $\hat{Z}_L^-$ is the partition function,

$$\hat{Z}_L^- = \sum_{\sigma} exp(-\beta \hat{H}_L^-(\sigma)) \tag{3.5}$$

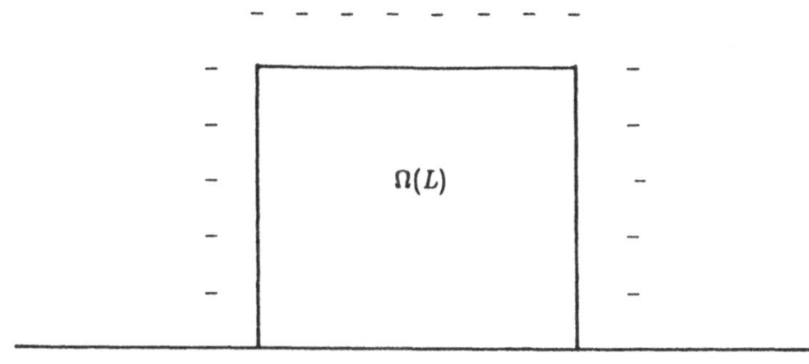

Fig. 3.

Any expectation values of local observables are computed with the probability measure ν_L^-. For example, the surface magnetization, per site, when we have the -phase in the bulk, is given at the thermodynamic limit by

$$m^-(\beta,h) = lim_{L\to\infty}(\sum_\sigma \nu_L^-(\sigma)\sigma(0)) \tag{3.6}$$

$\sigma(0)$ being the spin at $i=(0,\cdots,0)$.

The surface free energy $\tau^-(\beta,h)$ is obtained from $ln\hat{Z}_L^-$ by subtracting the bulk free energy. In fact we are not interested to all the surface free energy but only to the contribution due to the presence of the wall. Thus we define

$$2\tau^-(\beta,h) = lim_{L\to\infty} -\frac{1}{\beta}\frac{1}{(2L+1)^d} ln\frac{(\hat{Z}_L^-)^2}{Z_L^-} \tag{3.7}$$

The division by Z_L^- (see 2.3) allows us to subtract the bulk term and all boundary terms which are not related to the presence of the wall. (The definition which we use here differs from the one used in [2] by a constant independent of h). A similar definition gives $\tau^+(\beta,h)$. The first basic result is a mathematical proof of (1.1),

$$|\tau^+(\beta,h) - \tau^-(\beta,h)| \le \tau^\pm(\beta) \tag{3.8}$$

From this formula and from (2.12) we get immediately that

$$\tau^+(\beta,h) = \tau^-(\beta,h) \tag{3.9}$$

for all $\beta \le \beta_c(d)$. Equality (3.9) reflects simply the fact that there is only one bulk phase under these conditions. Let us fix the bulk phase, say the -phase, by choosing the -boundary condition. For large enough β it is possible to show that $\tau^-(\beta,h)$ is an analytic function of h for all $h < h^*(\beta)$ where $h^*(\beta)$ is strictly positive. In particular $\tau^-(\beta,h)$

is analytic at $h = 0$ under this hypothesis. Analogous properties are valid for $\tau^+(\beta, h)$, since by symmetry

$$\tau^+(\beta,h) = \tau^-(\beta,-h) \tag{3.10}$$

Another very useful formula is

$$\tau^-(\beta,h) - \tau^+(\beta,h) = \int_0^h ds(m^+(\beta,s) - m^-(\beta,s))ds \tag{3.11}$$

which is valid for all $h \geq 0$. Here $m^-(\beta,s)$ is defined in (3.6) and $m^+(\beta,s)$ is defined similarly using the measure ν_L^+. Whenever $\tau^-(\beta,h)$ is differentiable in h we have

$$\frac{d\tau^-}{dh}(\beta,h) = -m^-(\beta,h) \tag{3.12}$$

The integrand in (3.11) is a positive and decreasing function. Therefore $\tau^-(\beta,h)-\tau^+(\beta,h)$ is a monotone increasing concave function of the positive variable h. Since we can prove that

$$\tau^-(\beta,h) - \tau^+(\beta,h) = \tau^+(\beta), h \geq J \tag{3.13}$$

we have the following situation.

We can unambiguously define

$$h_w(\beta) = min\{h : \tau^-(\beta,h) - \tau^+(\beta,h) = \tau^{\pm}(\beta)\}$$

At $h_w(\beta)$ occurs the wetting transition according to the discussion of section 1. For $h > h_w(\beta)$ the wall is wetted by the +phase completely, so that the -phase cannot reach anymore the wall. From (3.11) and the monotonicity and positivity properties of $m^+(\beta,h) - m^-(\beta,h)$ we get confirmation of this result, since we must have

$$m^+(\beta,h) = m^-(\beta,h), h > h_w(\beta) \tag{3.14}$$

This means that the surface magnetization is independent of the nature of the bulk phase. In fact we have a much stronger result : the two measures $\nu^+ = lim_{L\to\infty}\nu_L^+$ and $\nu^- = lim_{L\to\infty}\nu_L^-$, which describe all local equilibrium properties of the system near the wall and at the thermodynamic limit, are the same for all $h > h_w(\beta)$. Formula (3.13) gives an upper bound on $h_w(\beta)$. Using again (3.11) we get a lower bound on $h_w(\beta)$, which is very good for $\beta \gg 1$. Clearly $|m^+(\beta,s) - m^-(\beta,s)| \leq 2$. Thus

$$\tau^{\pm}(\beta) = \int_0^{h_w(\beta)} ds(m^+(\beta,s) - m^-(\beta,s))ds \leq 2h_w(\beta) \tag{3.15}$$

This leads to the inequalities

$$\frac{1}{2}\tau^{\pm}(\beta) \leq h_w(\beta) \leq J \tag{3.16}$$

These results are so far valid for any dimension $d \geq 2$. They are summarized in the phase diagram of figure 5.

In this figure the exact phase boundary is known only for $d = 2$.

We finish this section by discussing the magnetization profile in the two-dimensional case. Let

$$m_L^-(\beta, h, z) = \sum_\sigma \nu_L^-(\sigma)\sigma(i_z) \tag{3.17}$$

with $i_z = (0, z)$. Taking the thermodynamic limit we get

$$m^-(\beta, h, z) = \lim_{L\to\infty} m_L^-(\beta, h, z) \tag{3.18}$$

Finally let $\beta_w(h)$ be the solution of the equation $h_w(\beta) = h$. (In two-dimension we know that there is a unique solution for $0 \leq h < J$). Abraham was able to compute the magnetization profile in [1]. His main results are

A) $\beta < \beta_w(h)$ and $0 < h < J$

In this case we have the complete wetting regime. The quantity $m^-(\beta, h, z)$ is the local magnetization at a distance z of a wall adsorbing preferentially the +phase. We have

$$m^-(\beta, h, z) = m^+(\beta, h, z)$$

and

$$lim_{z\to\infty} m^-(\beta, h, z) = m^*(\beta) \tag{3.19}$$

indicating that we are in the +phase of the system although we have used the -boundary condition. This result does not mean that the bulk phase is not the -phase. It shows only that the film of the +phase, which wets the wall, has an infinite thickness with respect to the unit lattice spacing . Indeed, Abraham was able to show that the magnetization $-m^*(\beta)$ of the -phase is attained by scaling z with L :

$$lim_{L\to\infty} m_L^-(\beta, h, \alpha L^\delta) = \begin{cases} -m^*(\beta), & \delta > \frac{1}{2} \\ +m^*(\beta), & \delta < \frac{1}{2} \end{cases} \tag{3.20}$$

This remarkable result shows in a precise way that the interface between the -phase in the bulk and the wetting film is typically at a distance of order $\sqrt{L}$ where L is the size of the box $\Omega(L)$. For the case $\delta = \frac{1}{2}$ see [1].

B) $\beta > \beta_w(h)$ and $0 < h < J$

In this case we are in the partial wetting regime. The -bulk phase reaches the wall. The results are completely different :

$$m^-(\beta, h, z) \neq m^+(\beta, h, z)$$

and

$$lim_{z\to\infty} m^-(\beta, h, z) = -m^*(\beta) \tag{3.21}$$

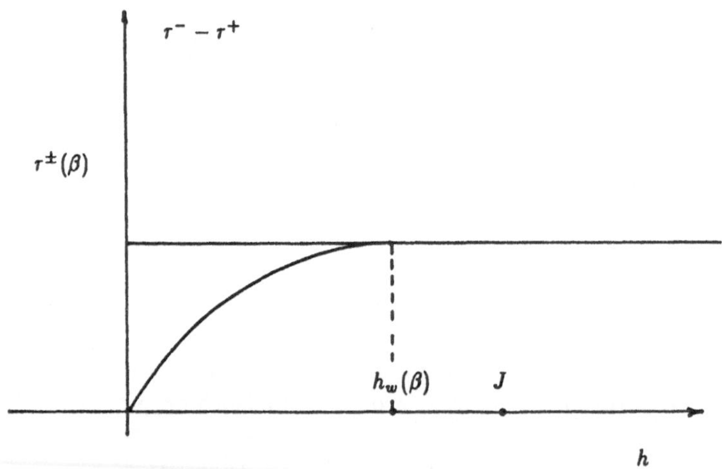

Fig. 4. β is fixed, $\beta > \beta_c(d)$

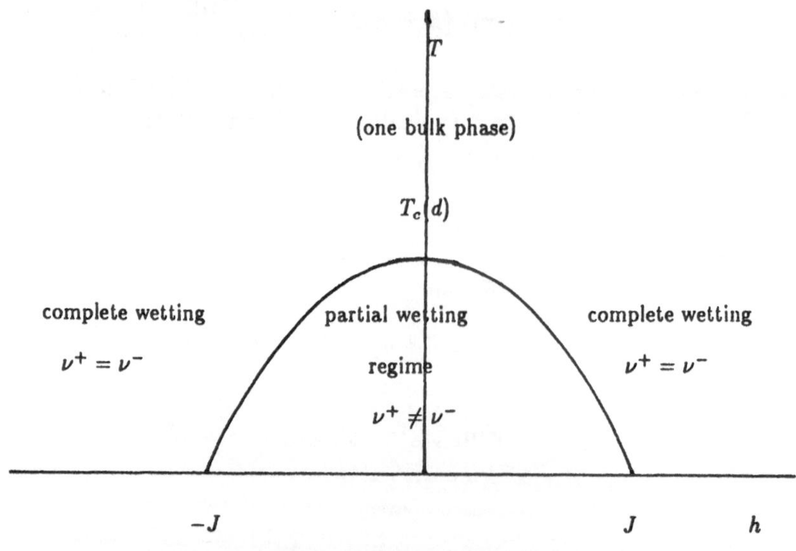

Fig. 5. Phase diagram

References

1. Abraham D.B. : Solvable model with a roughening transition for a planar Ising ferromagnet, Phys. Rev. Lett. 44 1165-1168 (1980).
2. Fröhlich J., Pfister C.E. : Semi-infinite Ising model I and II, Commun. Math. Phys. 109, 493-523 (1987) and 112, 51-74 (1987); The wetting and layering transitions in the half-infinite Ising model Europhys. Lett. 3 845-852 (1987).
3. Pfister C.E., Penrose O. : Analyticity properties of the surface free energy of the Ising model, Commun. Math. Phys. 115, 691-699 (1988).
4. De Coninck J., Dunlop F. : Partial to complete wetting : A microscopic derivation of the Young relation, J. Stat. Phys. 47 827-849 (1987).
5. Abraham D.B. : Surface structures and phase transition-exact results. In : Phase transitions and critical phenomena vol. 10 Domb, C., Lebowitz J.L. (eds) Academic Press (1986).
6. Shlosman S.B., : Wulff Construction Justified to appear in the proceedings of the 9th international congress on mathematical physics (1988).
7. De Coninck J., Dunlop F., Rivasseau V. : On the microscopic validity of the Wulff Construction and of the generalized Young equation, Commun. Math. Phys. 121, 401-419 (1989).
8. Gallavotti G. : Instabilities and phase transitions in the Ising model. A review. Riv. Nuovo Cimento 2, 133-169 (1972).
9. Pfister C.E. : Interface and Surface tension in Ising models. In : Scaling and self-similarity in physics. Fröhlich J. (ed.) Birkhauser (1983).
10. Lebowitz J.L., Pfister C.E. : Surface tension and phase coexistence, Phys. Rev. Lett. 46 1031-1033 (1981).

WETTING AT NANOSCOPIC SCALES: SOME EXPERIMENTS

F. Heslot, N. Fraysse, A.M. Cazabat, P. Levinson, P. Carles

Collège de France / Physique de la Matière Condensée.
11 Pl. M. Berthelot. 75231 Paris Cedex 05, France.

Abstract

This paper first addresses some aspects of wetting, when considered at a "microscopic scale", 0 to 1000 Å. In particular, the contact angle loses meaning at such scale. In a second part, recent experiments are presented, demonstrating that phenomena in the "nanoscopic range" (say 0 to 20 Å from the substrate) may play a dominant role although they are still poorly understood: Short-range forces and molecular dynamics are necessary to account for the spreading behaviour at such scale.

1: General introduction

Let us consider a droplet of fluid deposited on a flat horizontal smooth substrate. At a "macroscopic" scale (thickness $\geq 1\ \mu m$), it is usual to describe the edge of the drop by a contact angle θ . It has been known for a long time that a static contact angle reflects "interfacial contributions", respectively of the liquid-gas, liquid-solid, and solid-gas interfaces (Young 1805). In particular, it does not vary as the thickness of the solid varies from microns to millimeters, as was pointed out by J.C. Maxwell.

The question [1] we shall address is the following: How is the edge of a liquid droplet behaving, when examined at a "microscopic" scale (0 to 1000Å) ?

At this level, it becomes necessary to include the distance dependency of the various interactions: The attraction (or repulsion) of a molecule of fluid is expected to be stronger very close to the solid substrate (down to tens of angstroms, the order of a few typical bond lengths). This region will be designated as the "nanoscopic" range. The interactions decay at intermediate distance (typically 100 Å to 1000Å, typical order of magnitude for Van der Waals interactions, etc...). But it is only "far away" ($\geq 1\ \mu m$) that additional fluid thickness will not make a change in the total energy of interaction. It is thus clear that at the microscopic scale (0 to 1000Å), the thickness profile at the edge of a droplet (at equilibrium or spreading) will be strongly dependent of the nature of the forces involved. In particular, the contact angle is no more defined at such scale.

A usual description of the thickness dependence of the interactions consists in introducing [2] the "disjoining pressure". Let us denote by E(z) the free energy per unit surface of the solid covered by a slab of liquid of thickness z. The disjoining pressure is defined as $\Pi(z) = - dE/dz$.

This quantity is measurable, at least in a limited range, for example with the elegant experiment of the Russian school of Deryagin: The solid surface is immersed into a given wetting liquid, and a gas bubble is pressed against the surface. The liquid may resist to the thinning: For a given pressure P_0, one measures the thickness z_0, and one has then $\Pi(z_0)=P_0$. The measurement is feasible in the typical range 100 to 500 Å. It is then possible for example to test for Van der Waals interaction, but not all the "nanoscopic" range (short range forces). One may plot the expected features of the curve E(z) vs z.

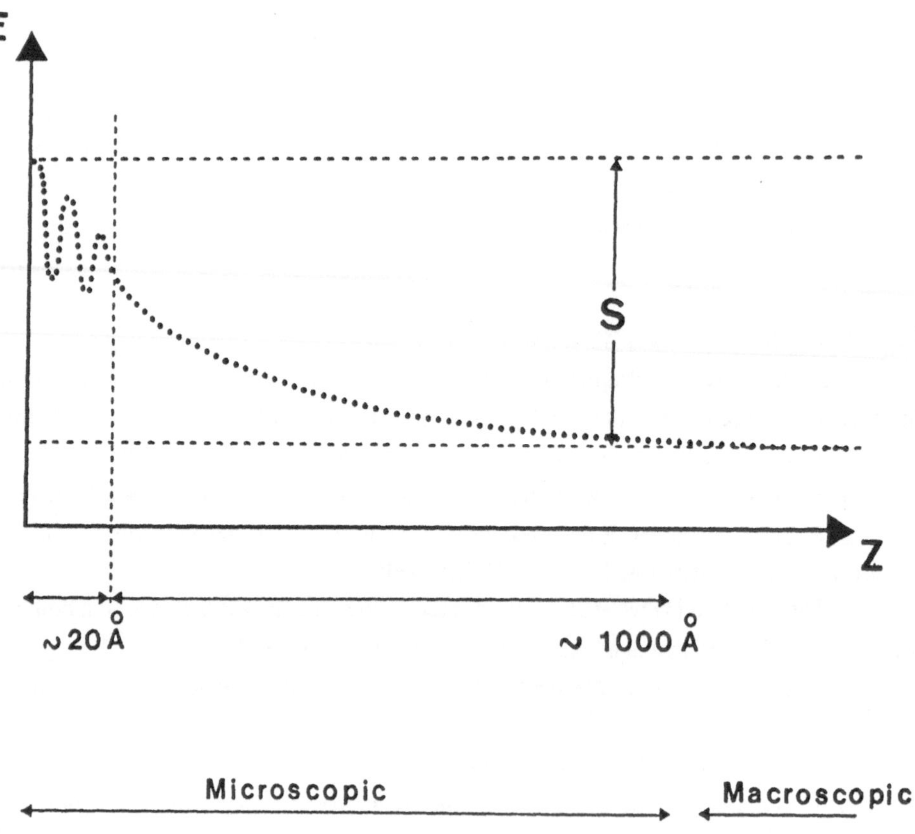

On the plot, two regimes are represented: The "long range" region (20 to 1000Å), and the "short range" region, in general not precisely known, but possibly with oscillations [3]. The parameter S that appears on the figure is the difference in free energy per unit surface between the situation of the bare solid in vacuum, and the solid covered with an infinite layer of liquid. The case S>0 corresponds to a "completely wetting" fluid, at the macroscopic scale. It is worth noting that even for S<0, ("non-wetting" at the macroscopic scale), there could be a thin wetting film, due for example to an oscillation of the short-range part of E(z).

In particular, the case of "autophobic" liquids is well known (term coined by Zisman [4]: Such a liquid does not wet a monolayer of itself).

Before proceeding to experiments in that range, let us recall some known experimental facts about the macroscopic level (contact angle). It is empirically known that solid surfaces may be separated into two loose categories: "low energy" surfaces (e.g. Teflon) , and "high energy" surfaces (clean metal oxide, clean glass...). But a frustrating experimental point is that the free energy per unit surface of the solid-liquid and solid-gas interfaces (γ_{sl} and γ_{sg}) are hardly totally measurable: We will recall essentially three techniques; one is the measurement of the heat of immersion (see for example [5]). It requires divided solids with large surface and may be corrupted by grain-size effects. The second, due to Zisman [4], is clever but empirical: It is experimentally shown in some specific cases that it is possible to define a value for the surface energy of a solid, with respect to an homologous series of liquids (e.g. alcanes of varying chain length). But this type of measurement is still restricted to homologous liquids, and to surfaces of "low" energy, where finite equilibrium contact angles are observable. The third technique could be derived from atomic forces measurements between two plates of mica with a liquid in-between [3]. But although one may be able to investigate the beginning of the short-range part (and in particular observe oscillations in some cases), questions still arise from the fact that two solid surfaces are involved.

2 : Some recent experiments on the "nanoscopic" range.

It has been known for a long time that an invisible ($< 1\mu m$) film of liquid, referred to as a "precursor film", precedes the spreading of a macroscopic droplet of a completely wetting fluid [6]. The field has been the subject of a number of theoretical [7,8] and experimental [9-17] studies. It is possible to study those films by grazing incidence X-ray reflectometry, or by ellipsometry. We will present recent experimental [15-17] ellipsometric studies on the time-dependent thickness profile of tiny droplets of fluid spreading completely on a smooth surface (silicon wafers).

Ellipsometry is a well known optical technique [18,19], which is sensitive to the presence of "submolecular films"tical technique [18,19], which is sensitive to the If the substrate is bare, incident light upon reflection at the Brewster angle is totally polarized (Fresnel laws). If a film of different refractive index is present at the surface of the substrate, the same experiment does not yield anymore total polarization. Inversely, the measurement of the change of polarization state upon reflection gives informations that may be interpreted for a film covered substrate, as the optical thickness of the covering film (product n . e, e= thickness, n= index of refraction), in the case $e \ll \lambda$ (λ= wavelength of light).

Our instrument uses an He-Ne laser and an acousto-optic modulation [20,21].The spatial resolution is 30 µm by 120 µm. The thickness measurement stability is of the order of 0.2 Å over one day. The substrates are (111) polished silicon wafers (index of refraction

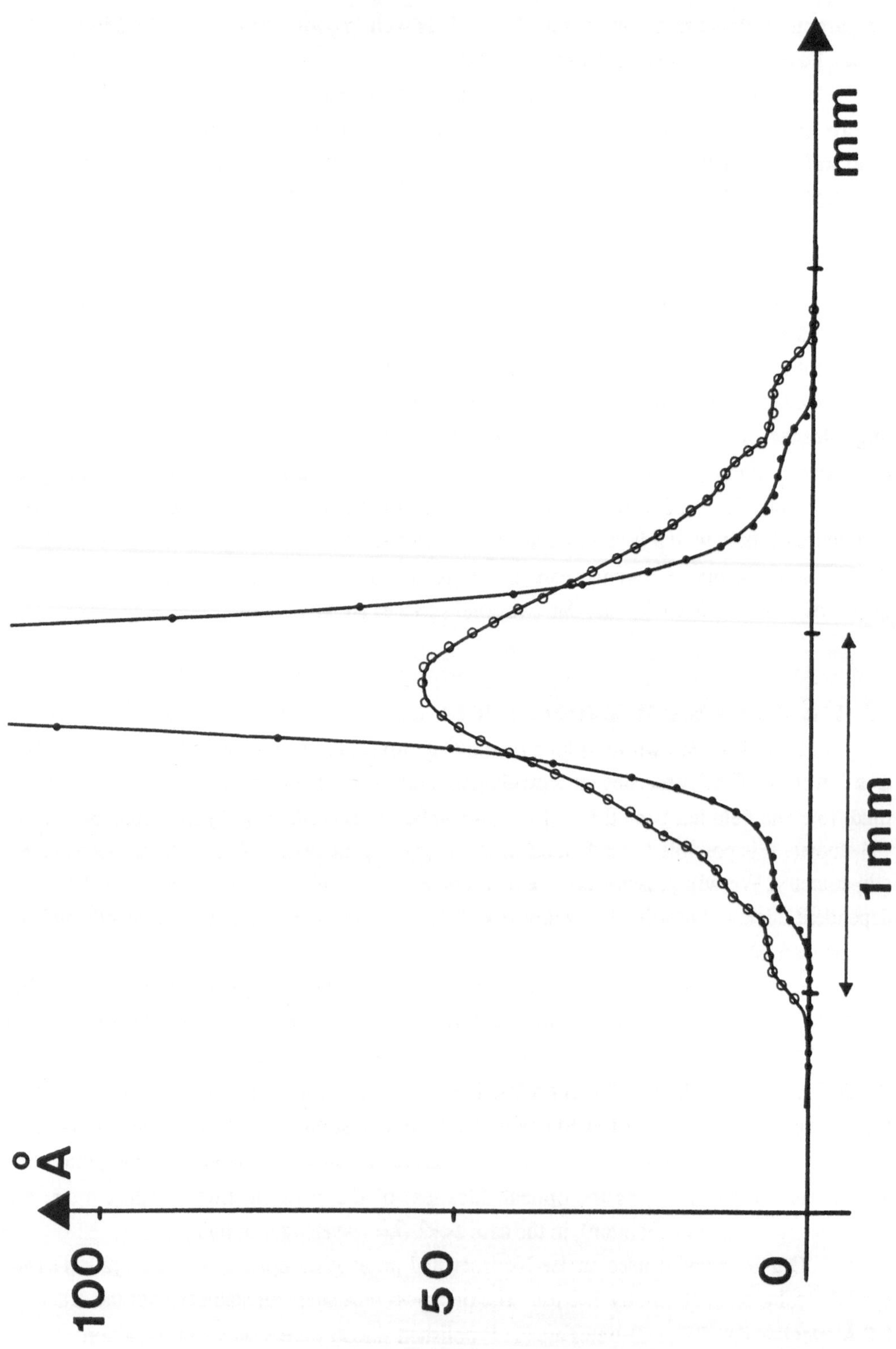

100
50
0
Å
1 mm
mm

n= 3.88-j 0.02, at 6328 Å). They are covered with natural oxide, whose growth is negligible over the experiment time (one week). The ellipsometric measurement takes into account both the oxide layer(index of refraction n= 1.4), plus the spreading fluid layer (typically n=1.4 to 1.5), plus possibly contamination. A base-line thickness profile is measured before depositing any oil, and subsequently substracted.

We have taken a great care to reduce contamination. The sample holder is in teflon, and is placed in a chamber at atmospheric pressure and room temperature, under a constant flow of filtered dry Nitrogen. It is possible to monitor the residual contamination by the measurement of the apparent thickness increase when no fluid is present. It was found to be typically 0.3 Å per week.

The roughness of the substrate is not directly known, but the following results emerge for this kind of substrate: Atomic force microscopy on (111) oxide covered silicon wafers, demonstrate [22] atomic smoothness over the scanned region (200 Å by 200 Å), with a few steps of about ten angstroms. The measured thickness is not affected by this residual roughness (linearity of the measurement at small thickness). It may nevertheless play a role in the film dynamics.

Following the loose distinction introduced in the first part of this paper, these oxide covered silicon surfaces are "high energy" surfaces. The fluid used is PDMS (molecular mass 2400, viscosity η=20 10^{-3} Pa.s, surface tension γ = 20.6 10^{-3}Nm^{-1}. The polydispersity index is 1.7 and no secondary peak is observed in the mass distribution [G.P.C. results. Courtesy of S. Boileau]). A tiny droplet is deposited on the surface of the substrate. Typical thickness profiles (measured 45 minutes, and 180 minutes after deposition) appear on next page. The vertical axis corresponds to thickness (angstroms), and the horizontal axis to a radial position (in millimeters) across the drop. The reader should keep in mind the enormous distorsion of scales.

On the first profile (τ = 45 mn.), the central part of the droplet is out of the vertical range. On the sides of the droplet, "feet" protrude. Their thickness is of the order of 6 to 7 Å, and theysides of the droplet, "feet" protrude. Their thickness is of the order of 6 to 7 Å, a diffusive process [17] , as long as the central part of the drop acts as a reservoir.This foot thickness is of the same order than the "lateral thickness" of a monomer [3], and is interpreted as a near-molecular carpet: The molecules are litteraly lying flat on the substrate.

On the second profile (τ= 180 mn), the drop has flattened below 70 Å, and is extending sideways. Three steps are now observable (the third one very faintly): The first step is advancing, followed by a second step (also advancing, but more slowly), and is then followed by a third step (advancing even more slowly), ... It is tempting to consider that the first layer coverage leaves a surface with a lower surface energy. This surface is still wettable by a second layer, and so on ..., leading to a kind of hierarchical process. But this model neglects possible molecular exchange between layers.

The end of the observed spreading process goes as follows: The upper layers retracts, and empty themselves into the lower ones; All the liquid is then found in the first layer, of near molecular thickness. This last layer is then slowly desagregating with time, by a diffusive process. This last evolution may depend both on surface heterogeneity, and molecular diffusion [15,17,23].

At the scale observed for all this spreading process, it is not possible anymore to use fluid dynamics to describe the flow, nor to restrict oneself to only Van der Waals forces for the interaction fluid-solid. It is necessary to consider the action of short-range forces and/or molecular dynamics. The dynamics of the advancing film [17], and the striking dynamic molecular structuration [18] may prove to be quite interesting theoretically (dynamic "nematic order " induced in the spreading fluid by the presence of a solid wall, adsorption/desorption dynamics, and for polymers possible effects of reptation,...).

It is tempting to consider that the first layer coverage leaves a surface with a lower surface energy. This surface is still wettable by a second layer, and so on ..., leading to a kind of hierarchical process. But this model neglects possible molecular exchange between layers.

The end of the observed spreading process goes as follows: The upper layers retracts, and empty themselves into the lower ones; All the liquid is then found in the first layer, of near molecular thickness. This last layer is then slowly desagregating with time, by a diffusive process. This last evolution may depend both on surface heterogeneity, and molecular diffusion [15,17,23].

At the scale observed for all this spreading process, it is not possible anymore to use fluid dynamics to describe the flow, nor to restrict oneself to only Van der Waals forces for the interaction fluid-solid. It is necessary to consider the action of short-range forces and/or molecular dynamics. The dynamics of the advancing film [17], and the striking dynamic molecular structuration [18] may prove to be quite interesting theoretically (dynamic "nematic order " induced in the spreading fluid by the presence of a solid wall, adsorption/desorption dynamics, and for polymers possible effects of reptation,...).

Conclusions:

Those recent experimental results on the microscopic range of wetting, demonstrate unambiguously for a "high energy" surface, the effect of short-range forces and molecular dynamics. More quantitative measurements are strongly needed, but it is already hoped that theoretical work will be developed to understand such effects.

References

1. De Gennes, P.G. : Rev. Mod. Phys. 57 (1985) 827.
2. Derjaguin, B.V. Churaev, N. V. and Muller, V.M. : "Surface forces" (Consultant Bureau New York and London) 1987; Adv. in Colloid and Int. Sci. 28 (1988) 197 and references.
3. Horn, R.G., Israelachvili, J.N. and Kott, S.J. :Macromol 21 (1988) 2836, and references.
3. Zisman, W.A. : in "contact angle, wettability and adhesion" (Gould, R.F., editor), Adv.
in Chemistry series 43 (American Chemical Society) 1964.
5. Wade, W.H. and Hackerman, N. : in "contact angle, wettability and adhesion " (Gould,R.F., editor), Adv. in Chemistry series 43 (American Chemical Society) 1964.
6. Hardy, H.W. : Phil. Mag. 38 (1919) 49.
7. Teletzke, G.F. Thesis, University of Minnesota (1983).
8. Joanny, J.F. and De Gennes, P.G. : J. Phys. Paris 47 (1986) 121.
9. Bascom, W.D., Cottington, R.L. and Singleterry, C.R. : in "contact angle wettability
and adhesion" (Gould, R.F., editor), Adv. in Chemistry series 43 (American Chemical ociety) 1964.
and adhesion" (Gould, R.F., editor), Adv. in Chemistry series 43 (American Chemical Society) 1964.
10. Beaglehole, D.: Unpublished results 1984, J. Phys. Chem., 93 (1989) 893.
11. Ausseré, D., Picard, A.M. and Léger, L. : Phys Rev. Lett. 57 (1986) 2671.
12. Daillant, J., Benattar, J.J., Bosio, L. and Léger, L. : Europhys. Lett. 6 (1988) 431.
13. Léger, L., Erman, M., Guinet, A.M., Ausséré, D., Strazielle, G., Benattar,J.J., Rieutord, F., Daillant, J. and Bosio, L.: Revue de Physique Appliquée 23 (1988) 1047.
14. Léger, L., Erman, M., Guinet, A.M., Ausseré, D., Strazielle, G., Phys. Rev. Lett. 60 (1988) 2390.
15. Heslot, F., Cazabat, A.M. and Levinson, P. : Phys. Rev. Lett. 62 (1989) 1289.
16. Heslot, F., Fraysse, N. and Cazabat, A.M. : Nature, 338 (1989) 1289.
17. Heslot, F., Cazabat, A.M. and Fraysse, N.: To appear in J. of Phys. Cond. Matter.Liquids. (1989).
18. Drude, P. : " The Theory of optics" (Dover New York) 1959.
19. Azzam, R.M.A. and Bashara, N.M.: "Ellipsometry and polarized light" (North Holland, Amsterdam) 1977.

20. Beaglehole, D. : Physica 100B (1980) 163; Rev. Sci. Inst. 59 (1988) 2557.
21. Drevillon, B., Perrin, J.,Marbot, R., Violet, A., and Dalby, J.L., Rev.Sci.Instrum 53(7), 969 (1982).
23. Marty, O., Drake, B., Gould, S. and Hansma, P.K. : J. Vac. Sci. Technol. A6 (1988) , O., Drake, B., Gould, S. and Hansma, P.K. : J. Vac. Sci. Technol. A6
24. Pomeau, Y., C.R. Acad. Sci. 298 II (1983) 29.

ADSORPTION AND THEORY OF FLUIDS

J. Bougard and R. Jadot

Polytechnic Faculty of Mons, Department of Thermodynamics, 31, boulevard Dolez, 7000 MONS, Belgium

1. Introduction

The applications of adsorption phenomena are numerous: gas purification, gas extraction, gas storage, heterogeneous catalysis, refrigeration, heat storage, vacuum technology ...

The present paper concerns the analysis of experimental data obtained in the field of a research programme about solar refrigerating machines and heat storage. These machines are based on the adsorption of selected gases, generally named refrigerants, on adequate adsorbents. The working pressures range from vacuum to the saturation pressure and the working conditions often approach the critical point of the refrigerant.

In Table 1 are listed the different gas-adsorbent systems under consideration while Table 2 gives the main properties of the adsorbed gases.
The adsorbents are mainly characterized by their specific area based on the adsorption of nitrogen at low temperature and using the BET equation.
The value is 750 m^2/g for the Silicagel-KC and 1000-1200 m^2/g for the different activated charcoals.
More details may be found in the references.

For each system, the experimental measurements give the mass m of the adsorbed gas per mass unit of adsorbent as a function of the pressure and the temperature

$$m = f(T, P)$$

It is desired to select the most adequate equation able to correlate the data in the whole range of pressure and temperature.

In a first phase, the well known Langmuir's, B.E.T's and Dubinin's equations have been tested (1).

Two criteria are used to judge the validity of these equations:

- the mean absolute deviation:

$$OF1 = \frac{1}{N} \cdot \sum_{i=1}^{N} |m_{exp} - m_{cal}|$$

where N is the number of data, m_{exp} and m_{cal} the experimental and the calculated values of m.

- the mean relative absolute deviation

$$OF2 = \frac{1}{N} \cdot \sum_{i=1}^{N} |\frac{m_{exp} - m_{cal}}{m_{exp}}|$$

2. Langmuir's model [7]

The adsorbed gas is supposed to be distributed in a monolayer covering a well defined active surface of the adsorbent. The interactions between the adsorbed molecules are neglected.

The resulting equation is

$$\frac{m}{m_{max}} = \frac{bp}{1 + bp} \tag{1}$$

m_{max} is the maximum value of m corresponding to a complete covering of the surface. b is the Langmuir parameter, function of the temperature and of the nature of the adsorbate. Equation (1) may be rewritten as follows

$$\frac{p}{m} = \frac{p}{m_{max}} + \frac{1}{m_{max}.b} \tag{2}$$

showing that (p/m) is a linear function of p. This is generally not the case for our data as shown for instance on the figure 1.

Consequently, the deviations between theory and experience listed, in the Table 3, are large.

3 Brunauer-Emmet-Teller's model [7]

Following BET, the adsorbed phase consists of several layers starting from the pore surface. A strong interaction exists only between the surface and the first layer.

If n, the number of layers, is infinite, the BET model leads to

$$\frac{m}{m_m} = \frac{cx}{(1-x)(1-x+cx)} \tag{3}$$

$$x = p/p_s(T) \tag{4}$$

where m_m is the mass corresponding to a single layer c is a parameter and $p_s(T)$ the saturated vapour pressure of the adsorbate.

Table 1. Data collections

Gas	Adsorbant	Temperature °C	Number of data	Reference
R12 CF_2CL_2	CA-BPL (CA1)	25,50,80	8,8,7	1,2,3
R12 CF_2CL_2	SI-KC	30,40,60,80	9,13,11,7	1,2
R22 CHF_2CL	CA-BPL (CA1)	50	8	2,3
R22 CHF_2CL	SI-KC	25,40,60,80	15,10,7,5	1,2
R717 NH_3	CA-BPL (CA1)	30,60,90	9,7,9	1
R170 C_2H_6	CA-MSC-SA (CA2)	5,4,30,50	7,8,9	4
R290 C_3H_8	CA-MSC-SA (CA2)	5,4,30,50	9,9,7	4
CH_3OH	CA-NORTI (CA3)	20,40	11,11	6
C_2H_5OH	CA-B5/1,8 (CA4)	40	13	6
C_2H_5OH	CA-35/3 (CA5)	40	1.9	6
C_2H_5OH	CA-40/3 (CA6)	40	24	6
R12-R22	SI-KC	40	6	2
R12-R22	CA-BPL	40,50	12,21	2,3
R170-R290	CA-MSC-SA	5,4,30,50	20	5

Rewriting (3) as

$$\frac{p}{m(p_s(T)-p)} = \frac{1}{m_m c} + \frac{c-1}{m_m c}\left(\frac{p}{p_s(T)}\right) \tag{5}$$

shows that the function $p/m(p_s(T)-p)$ is a linear function of $x = p/p_s(T)$.

The agreement is generally poor as shown on figure 2.

However, recalling that the average diameter of the micropores is in the range 20-30 Å and that the diameters, of the gas molecules is in the range 3-6 Å it is not physically sound to imagine a infinite value for n.

For a number of layers finite, equation (3) becomes

$$\frac{m}{m_m} = \frac{cx}{1-x}\cdot\frac{1-(n+1)x^n + nx^{n+1}}{1+(c-1)x - cx^{n+1}} \tag{6}$$

Table 2. Properties of the adsorbed gases

	Molecular mass	Normal ebullition	Critical temperature T_c(K)	Critical pressure P_c(bar)	Critical volume (cm3/mol)	Acentric factor w
R12-CCl2F2	120.9	-29.8	385.15	41.33	216.7	0.177
R22-CHClF2	86.47	-40.8	369.15	49.77	164.74	0.218
R717-NH3	17.03	-33.35	405.55	112.97	72.55	0.251
R170-C2H6	30.06	-88.6	305.42	48.21	148.01	0.105
R290-C3H8	44.09	-42.2	369.79	42.01	210.09	0.152
CH3OH	32.12	64.6	513.20	78.55	118.11	0.557
C2H5OH	46.08	78.8	516.15	63.00	161.9	0.638

Table 3. Mean Absolute Deviation

System gas-adsorbent	Mean Absolute Deviation OF1 [10^{-3} kg/kg]			
	LANGMUIR	BET (finite n)	DUBININ	WILSON
R12-CA_1	13	5	40	5
R12-SI	55	11	24	16
R22-CA_1	10	4	21	5
R22-SI	63	10	18	12
R717-CA_1	43	9	10	7
R170-CA_2	8	9	2	2
R290-CA_2	6	16	2	3
CH_3OH-CA_3	30	8	7	6
C_2H_5OH-CA_4	4	3	6	7
C_2H_5OH-CA_5	6	5	14	5
C_2H_5OH-CA_6	5	3	13	6

Adjusting the two parameters c and n, (6) leads to a much better fitting between theory and experience as shown on figure 2 and table 3.

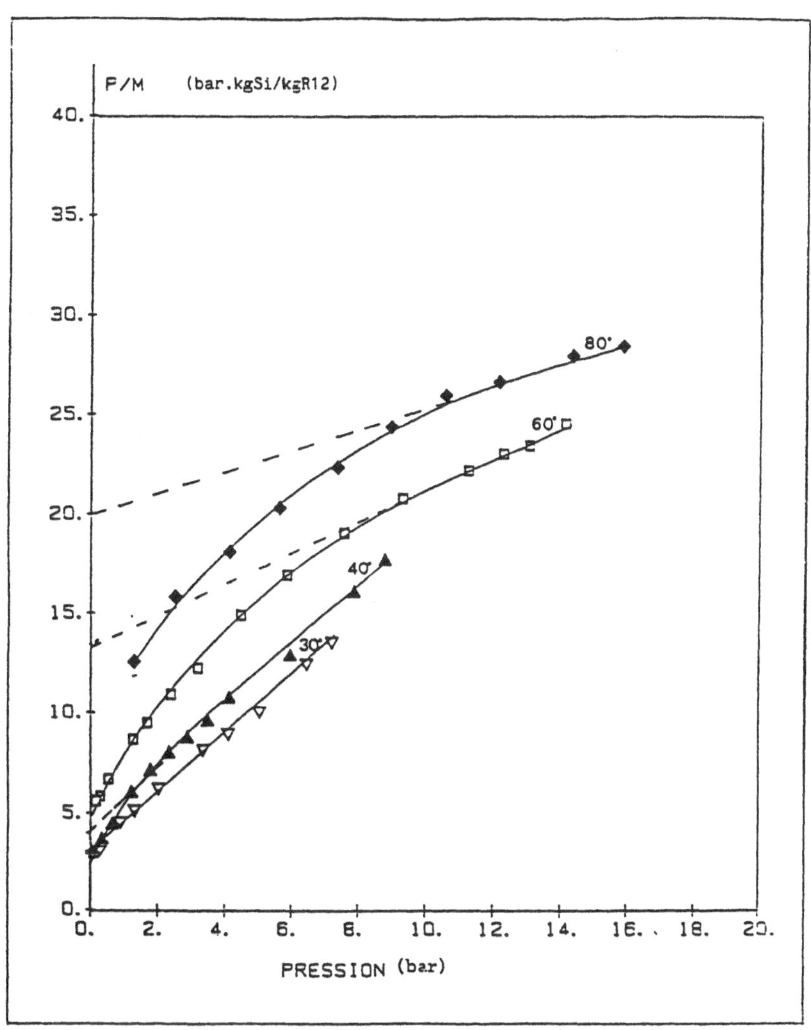

Fig. 1. Langmuir's equation. Adsorption of R12 on Silicagel KC

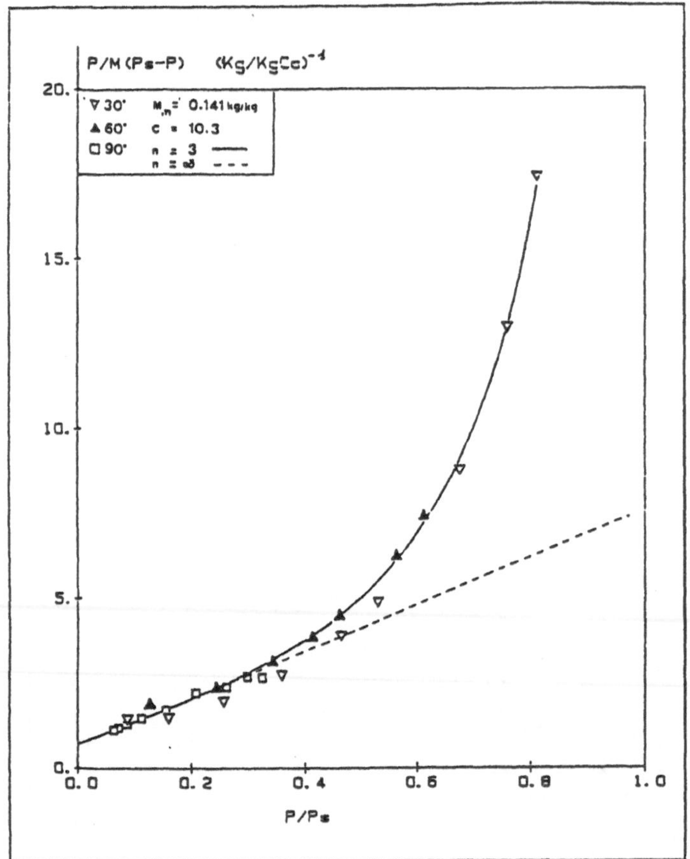

Fig. 2. Adsorption of R717 (NH_3) on activated charcoal BPL following BET graph.

4 Dubinin's model [8-9]

The adsorbate fills the micropores starting from the surface. The volume V of the adsorbed phase (figure 3) is a function of the Polanyi potential, A=RT ln (p/ps(T)), independent of T for a given adsorbate-adsorbent pair. The distribution of the volumes of the micropores is assumed to be gaussian.

The Dubinin's equation is

$$logm = log(V_0 \rho_L) - D(T log \frac{p_s}{p})^2 \tag{7}$$

where V_0 is the total volume of the micropores, ρ_L is the density of the adsorbed phase that assumed to be equal to that of the liquid phase and D is the Dubinin's parameter.

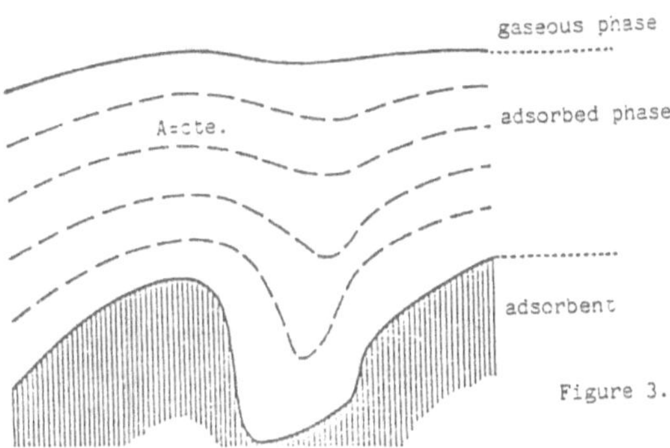

Fig. 3.

The linear form of (7) permits an easy adjustment of the parameters V_0 and D for a given gas-solid pair.

Figure 4 shows that the linearity of (7) is not always verified specially close to saturation.

It has been suggested (10) to adjust the exponent in the last term of equation (7). However for our data this does not improve the fitting.

Let is note that to take into account the imperfection of the gaseous phase the pressure has been replaced by the fugacity.

Table 3 shows that the predictions obtained with Dubinin's equation are not better than those obtained with BET's equation.

5 Thermodynamical approach

In the adsorbed phase, the component, named 1, coming from the gaseous phase exists in a condensed state at a pressure p lower than the saturation pressure $p_s(T)$.

This behaviour is similar to that of the component 1 in a liquid solution containing another component 2. In this case, the interactions between 1 and 2 leads to a reduction of the chemical potentiel of 1 allowing thus a reduction of the equilibrium pressure. In the adsorbed phase, the reduction is due to the strong interactions between 1 and the solid adsorbent.

Adsorption may thus be considered as the equilibrium between a pure gaseous phase of component 1 at (T,p) and a liquid or solid mixture of components 1 and 2.

Thermodynamics does not need at all to know what component 2 exactly is. However, 2 may be understood as the support of the influence of the solid on the properties of the adsorbed gas 1.

Using the general formalism of chemical thermodynamics (11), one obtains

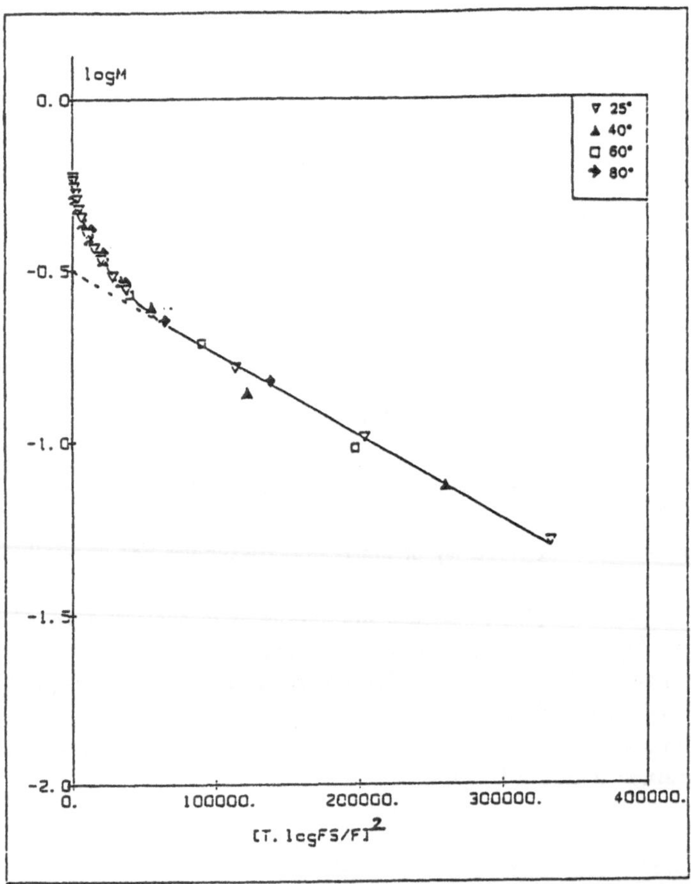

Fig. 4. Dubinin's equation. Adsorption of R22 on silicagel KC.

$$f_1 = fs_1 \frac{m_1}{m_{1max}} . \gamma_1 \tag{8}$$

$$\lim_{m_1 \to m_{max}} . \gamma_1 = 1 \tag{9}$$

where f_1 is the fugacity of the gas, f_{s1} the fugacity under saturated conditions, m_1 and m_{1max} the mass of the adsorbed gas at (T,p) and its maximum value respectively and γ_1 the activity coefficient of 1 in the solution (1+2).

At low pressure, (8) reduces to

$$p = p_s(T) . \frac{m}{m_{max}} . \gamma_1 \tag{10}$$

The activity coefficient is a function of the temperature of the mass fraction m/m_{max} and of the solution model.

WILSON's model (11) which takes into account a non random distribution of the components due to different and strong interactions is adequate for this problem. γ_1 is given by the following relation containing only two parameters Λ_{12} and Λ_{21}

$$ln\gamma_1 = -ln(x_1 + \Lambda_{12}x_2) + x_2[\frac{\Lambda_{12}}{x_1 + x_2\Lambda_{12}} - \frac{\Lambda_{21}}{x_2 + x_1\Lambda_{21}}]$$

$$x_1 = \frac{m}{m_{max}} \qquad x_2 = 1 - x_1$$

$$\Lambda_{12} = \frac{V_2}{V_1}exp(-\frac{\lambda_{12} - \lambda_{11}}{RT}$$

$$\Lambda_{21} = \frac{V_1}{V_2}exp(-\frac{\lambda_{12} - \lambda_{22}}{RT})$$

$(\lambda_{12} - \lambda_{11})$ et $(\lambda_{12} - \lambda_{22})$ are constant with temperature.

Equations (10 and (11) give thus the adsorbed mass $m(T,p)$ using only three parameters $(\lambda_{12} - \lambda_{11})$, $(\lambda_{12} - \lambda_{22})$ and m_{max}

The results are very satisfactory.

Table 4 shows the values of the mean relative deviations (OF2) using Wilson's model and Dubinin's equation.

Table 4. Comparison between Wilson's and Dubinin's relations.

System	Wilson		Dubinin
	V_0(cm^3/kg) at 50°C	OF2(%)	OF2(%)
R12-CA1	0.443	0.9	7.6
R12-SI	0.510	3.0	4.6
R22-CA1	0.449	1.0	8.0
R22-SI	0.503	2.0	6.4
R717-CA1	0.451	3.4	4.0
R170-CA2	0.188	3.3	3.4
R290-CA2	0.187	3.7	3.2
CH3OH-CA3	0.530	1.5	1.7
C2H5OH-CA4	0.612	1.4	1.5
C2H5OH-CA5	0.489	1.7	4.0
C2H5OH-CA6	0.487	2.2	5.0
R12-R22-CA1		6.8	impossible
R170-R290-CA2		9.8	impossible
OF2 = mean relative absolute deviation.			

The thermodynamical approach increases largely the accuracy.

The method gives also the volume V_0 of the micropores which is assumed to be equal to v_2.

It is important to observe in table 4 that the data of adsorption of different gases on the same adsorbent give the same value of V_0 showing therefore the consistency of the method.

6. Adsorption of mixtures.

The WILSON's model of multicomponent mixtures needs only binary parameters Λ_{ij}. The adsorption of a gas mixtures 1-3, for instance R_{12}, R_{22}, on the activated charcoal CA1 named 2 may thus be predicted using the same formalism.

The six parameters Λ_{ij} are obtained as follows

Λ_{12} and Λ_{21} from adsorption data of R_{12} on CA_1 [1]

Λ_{32} and Λ_{23} from adsorption data of R_{22} on CA_1 [1]

Λ_{12} and Λ_{31} from liquid-vapour equilibrium data [12]

Table 4 shows that the first results are promising.

References

1 Mahamane, A. Etude de l'adsorption de vapeurs sur solides microporeux. Adaptation de la Théorie des solutions,thesis,Université de Mons,1987.

2 Desondre, C. Adaptation de la théorie des solutions aux phénomènes d'adsorption. T.F.E., Fac. Pol. Mons, 1987.

3 Berteau, Ph. Etude de l'adsorption de vapeurs sur solides poreux. Adaptation de la théorie des solutions. T.F.E., Fac. pol. Mons, 1988.

4 Nakahara, T., Hirata, M., Omori, T. Adsorption of hydrocarbons on carbon molecular sieve. J. Chem. Eng. Data, 1974, vol. 19, n° 4, 310-313, 1974.

5 Nakahara, T., Hirata, M., Komatsu, S. Adsorption of a gaseous mixture of ethane and propane on a carbon molecular sieve.
J. Chem. Eng. Data, 1981, vol. 26, n°, 161-163, 1981.

6 Boussehain, R., Feidt, M.L. Caractérisation thermostatique du couple charbon actif-alcool. Incidence sur les critères de choix associés aux machines à quatre sources de chaleur. Thèse, Nancy, 1985.

7 Brunauer, S., Copeland, L.E., Kantro, D.L. The Langmuir and BET Theories. The solid gas interface. Vol. 1, Marcel Dekker Inc., New-York, 1967.

8 Bering, B.P., Dubinin, M.M., Serpinsky, V.V. Theory of volume filling for vapor adsorption. J. Colloid. Interf. Sci., vol. 21, 378-393, 1966.

9 Dubinin, M.M. Adsorption in micropores. J. Colloïd Interf. Sci., 23, 487-499, 1967.

10 Dubinin, M.M., Asthakov, V.A.. Adv. Chem. Serie, 102, 69, 1970.

11 Prausnitz, J.M. Molecular Thermodynamics of fluid phase equilibria. Pretice Hill, Inc., 1969.

12 Loffler, H.J. Eininge ergenschaften des binären systems frigen 12-frigen 22 der ternären systems F12-F22 - Naphthenbasisches Mineralöl. Kaltetechnik, 12, heft 9, 1960.

FRONT PROPAGATION IN ONE DIMENSION

P.Collet

Centre de Physique Théorique*, Ecole Polytechnique, F-91128 Palaiseau Cedex (France)

1 Introduction

A very common problem for systems out of equilibrium can be formulated as follows. Suppose a substance can exist in two phases. Assume an initial condition has been prepared such that the two phases coexist in different regions of space with a common boundary (interface). What is the future evolution for the system? This problem arises for example in solidification (dendrites, spinodal decompositions and droplets, etc.), but also in hydrodynamics (spatial intermittency), chemistry, biology etc. If one considers a sharp interface one is lead to a moving boundary problem. But very often the boundary is not sharp (consider for example temperature fields, concentration fields etc.). The physical process driving the time evolution of the system is usually described by dissipative (diffusive) nonlinear partial differential equations (the Navier-Stokes equation is a typical example of non linearity). These equations are of parabolic type and we shall assume that they define a well behaved semi group of time evolutions on a space of regular and bounded functions. Since we are working with infinite spatial domains this result may not be so easy to prove. We refer to Collet-Eckmann 1989 for a general introduction to this class of problems.

Up to now the problem has been mathematically analysed mostly in one dimension (except for the recent results on dendrites and the Saffman-Taylor problem). In space dimension larger than one, a major open problem is to understand the instabilities of the interface which can probably become fractal. In the sequel I will only discuss one dimensional problems.

In order to fix notations and give a more precise formulation of the problem, one can consider a list of standard equations of increasing complexity (and with a decreasing amount of results). The equations below describe the time evolution of a continuous field u which is for example a temperature or concentration field. The simplest equation is the real amplitude equation (or real Newell-Whitehead equation) subsequently referred by RNW. It acts on a real field u and is given by

* Laboratoire CNRS UPRA 14.

$$\dot{u} = u'' + u - u^3 , \qquad (RNW)$$

where $\dot{}$ denotes the time derivative and $'$ the space derivative. A more general situation including in particular the family of equations

$$\dot{u} = u'' + f(u) ,$$

where f is differentiable, and $f(0) = f(1) = 0$ with $f \geq 0$ on $[0,1]$ has been extensively discussed by Aronson and Weinberger (Aronson-Weinberger 1978) and also Bramson (Bramson 1983), mostly for the case of non negative solutions. We shall come back to their analysis later on.

The next level of complexity is given by the so called amplitude equation, also called Newell-Whitehead, or Ginzburg-Landau equation. We shall denote it by CNW for short. The CNW acts on a complex field v and is given by

$$\dot{v} = v'' + v - v|v|^2 . \qquad (CNW)$$

This equation is in some sense a normal form for a bifurcation of a continuous spectrum and this explains why it is so often used as a physical model.

The next stage of complexity is given by the Swift-Hohenberg equation (referred as the SH equation below). It was obtained as an approximation for the Bousseinesq system and the two dimensional version reproduces qualitatively the experimental behaviour. It acts on a real field and is given by

$$\dot{u} = \epsilon u - (1 + \partial_x^2)^2 u - u^3 , \qquad (SH)$$

where ϵ is a real parameter.

The above list can of course be continued by adding many other equations (Cahn-Hilliard, Navier-Stokes etc.). In some situations one has to add a noise term to model fluctuations (see for example Landau-Lifshitz 1959). We shall not consider these problems here.

As explained above, we shall investigate the question for time evolution of an initial condition describing the coexistence of two different phases. We shall adopt here a rather conservative definition of phases. Namely, we shall call a phase any stationary solution of the evolution equation. For the RNW the stationary solutions $u = 0$ and $u = \pm 1$ (among others) are phases. There is a noticeable difference between them. The solution $u = 0$ is linearly unstable, while each solution $u = \pm 1$ is linearly stable. Therefore we expect that an initial condition $u(x,0)$ interpolating between 1 and 0 ($u(-\infty,0) = 1$ and $u(+\infty,0) = 0$) for example will evolve toward the constant solution 1. In other words the domain occupied by the more stable phase will grow at the expense of the domain occupied by the least stable one. The main problem is to understand how this happens.

2 Fronts for the amplitude equation

An important notion in this problem is the concept of a front. We shall first explain it for the case of phases which are constant functions (as above) and give later on a more general definition including the case of non spatially homogeneous phases. Intuitively, a front is a special solution of the dynamical equation which interpolates between the two phases, travels at a fixed speed with a constant profile in the moving frame. It is therefore quite analogous to a soliton except that we are considering dissipative situations. More precisely, a front of the RNW from the phase 0 to the phase 1 at speed $c \geq 0$ will be a function l_c of a real variable such that

$$\begin{cases} l_c(-\infty) = 1\,, \\ l_c(+\infty) = 0\,, \\ l_c(x - ct) \text{ is a solution of the RNW.} \end{cases}$$

It is easy to verify that l_c will be a front at speed c if it satisfies the equation

$$l_c'' + cl_c' + l_c - l_c^3 = 0$$

with the boundary conditions $l_c(-\infty) = 1$ and $l_c(+\infty) = 0$. This equation is easy to discuss because it corresponds to a mechanical pendulum with hamiltonian $H = p^2/2 + q^2/2 - q^4/4$ and friction coefficient c. One can summarise the results in the following proposition (see Aronson-Weinberger 1978 for a more general discussion).

Proposition 1. *For each positive speed c there is a real front l_c from 0 to 1 at speed c. This front is positive for $c \geq 2$. Moreover, these fronts decay exponentially fast at $+\infty$ with an exponential rate γ_c given by*

$$\gamma_c = \frac{c - \sqrt{c^2 - 4}}{2}\,.$$

We first observe that although the equation is of diffusive nature, the non linearity was able to produce a non zero speed instead of a zero speed diffusive behaviour for the linear diffusion equation. We also note that there is a front for any speed. However the experiments show a definite and very reproducible speed. This raises the question of a physical selection mechanism for the speed.

We shall now explain briefly the deep results of Aronson and Weinberger which determine the asymptotic speed of propagation of a positive initial condition interpolating between 0 and 1. We shall formulate this result as follows (Aronson-Weinberger 1978, Bramson 1983).

Theorem 2. *Let $u(x, 0)$ be a regular positive initial condition which converges to 1 for $x \to +\infty$, and $u(x, 0) \sim \exp(-\gamma x)$ for large x (we refer to the original papers for a precise definition of this equivalence). Let c be defined by*

$$\begin{cases} \gamma_c = \gamma \text{ if } \gamma < 2\,, \\ c = 2 \text{ otherwise.} \end{cases}$$

Then, if $u(x,t)$ *is the solution at time* t *with initial condition* $u(x,0)$ *at time* 0, *there is a function* $m(t)$ *such that*

$$\lim_{t\to\infty} m(t)/t = 0$$

and

$$\sup_{x\in\mathbf{R}} |u(x,t) - l_c(x - ct - m(t))| \xrightarrow[t\to\infty]{} 0$$

(see Bramson 1983 for logarithmic estimates on $m(t)$*).*

In other words, except for a shift, in the frame moving at the correct speed, the profile converges to the profile of the corresponding front. Moreover, if the decay of the initial condition is fast enough, the limiting speed is 2. We can say that 2 is the speed corresponding to the largest set of initial conditions.

This theorem can be extended to the more general non linearities mentioned above. If the non linearity is concave, the selected speed is always given by twice the square root of the slope of the nonlinearity at the origin. However this may not be true for non concave functions. An extreme situation is given by a nonlinearity with slope zero at the origin and which is not identically zero. Then one can show that the selected speed is strictly positive. However, no explicit expression has been derived up to now for this speed.

The above theorem mostly applies to positive solutions. The reason is that the proof is based on the parabolic maximum principle. There is no reason to expect an extension of the proof to the CNW which is not a real equation, or to the SH which is a fourth order equation. It is therefore necessary to give another selection criteria which can be applied to more general situations. As we shall see, this can be done in terms of a local stability analysis of the fronts, but we shall loose the globality of the theorem of Aronson-Weinberger.

The more general selection criteria is known as the marginal stability criteria. It is a minimal set of stability properties that should be satisfied to have selection. Note however that this criteria may not be strong enough in all cases. The main idea is to perform a linear stability analysis for the fronts in the moving frame. One should first state precisely the space of admissible perturbations because the spectrum of the linearized evolution may depend strongly of this space. By analogy with the result of Aronson-Weinberger we shall consider a family of spaces $\mathcal{B}_\gamma$ defined by

$$\mathcal{B}_\gamma = \{h \,|\, h(\cdot)(1 + e^{\gamma\cdot}) \in L^\infty\}\,.$$

This is a Banach space when equipped with the natural norm. As explained above, if we consider the stability of a front l_c of speed c in the moving frame, this spectrum will depend of the value of γ determining the space $\mathcal{B}_\gamma$ of allowed perturbations. If γ is too large, the front will be stable, and similarly for small γ the front will be unstable (see Sattinger 1977). A natural choice for γ is the value γ_c which gives the decay of the front. In $\mathcal{B}_{\gamma_c}$, we expect the spectrum to be marginally stable because of the mode corresponding to the translation of the front. In fact since we want to have this mode for physical reasons, we want to impose the condition $\gamma \leq \gamma_c$. The result of the spectral analysis can be formulated as follows.

Proposition 3. *For the RNW, if* $c < 2$ *the front* l_c *is unstable in* $\mathcal{B}_{\gamma_c}$. *If* $c \geq 2$, *the front* l_c *is marginally unstable in* $\mathcal{B}_{\gamma_c}$.

This result is therefore a partial answer to the selection problem: it excludes the slowest fronts because they are unstable in their moving frame. We also observe that this result is valid for fronts which are not positive while it does not discriminate among the fronts travelling at speed faster than 2. In order to complete the selection we need a more delicate argument which is based on convective instabilities.

Assume that we fix a speed $c > 2$. Let c' bet a speed larger than c. We consider an observer travelling at speed c'. We assume once for all that these observers start at a position which is very far away in front of the center of the front (defined for example by $l_c = 1/2$). Assume now we perturb the front l_c by an initially small element of $\mathcal{B}_\gamma$, with $\gamma < \gamma_c$. We can ask if this perturbation will reach the observer or not. The result is summarized in the following proposition.

Proposition 4. *If c' is near enough to $c > 2$, for any γ near enough but smaller than γ_c, there are initially small perturbations in $\mathcal{B}_\gamma$ which will reach the observer travelling at speed c'.*

For $c = 2$, the situation is quite different. Namely, if we fix $c' > 2$, if γ is smaller than γ_c and near enough, an initially small perturbation in $\mathcal{B}_\gamma$ will not reach an observer travelling at speed c'.

This second stability result completes the marginal stability criteria for the RNW.

3 Fronts for more general equations

We shall now explain how the above last result can be extended to more general situations which do not satisfy the parabolic maximum principle. The first such case is given by the CNW. As before we have to identify the phases of this system. It is easy to verify that $v = 0$ is an unstable stationary solution. Moreover we have now a one parameter family of periodic stationary solutions given (except for a fixed constant phase by

$$v_q(x) = \sqrt{1 - q^2}e^{iqx} \quad \text{for} \quad -1 \le q \le 1\,.$$

It is not very difficult to show that these solutions are marginally stable if $|q| \le 1/\sqrt{3}$, and otherwise unstable. The values $q = \pm 1/\sqrt{3}$ correspond to the well known Eckhaus instability. Note that for $|q| \le 1/\sqrt{3}$ we cannot expect strict stability for two reasons. One is the existence of the continuous one parameter family of solutions. The other reason is the marginal mode associated to the translation of the above solutions.

We shall now investigate the problem of evolution between the unstable phase $v = 0$ and one of the phases v_q for a fixed q of modulus smaller than $1/\sqrt{3}$. The first question is the existence of fronts. It turns out that one can look for a front which is the stationary solution v_q multiplied by an envelope function moving at speed c. One finds an equation for the envelope and one can prove a result analogous to proposition 1. Namely for any positive speed c we have a front from 0 to v_q at speed c. One can then repeat the linear stability analysis of the fronts trying to apply the marginal stability criteria. The selection is then even deeper than expected. It is only for the frequency $q = 0$ that one finds a front satisfying the above marginal stability criteria. In other words, we also have here a pattern selection because the only stationary marginally stable solution that can emerge from the unstable solution $v = 0$ via a front which satisfies the criterion is the uniform solution. This solution corresponds to the real solution treated in the RNW and the marginal speed is again 2.

We now consider the more difficult case of the SH equation. It is more difficult than the CNW because as we shall see it is not possible to look for fronts which are given by a stationary solution multiplied by a moving amplitude.

Here again we start the analysis by looking for stationary solutions. We can use the parameter ϵ as a bifurcation parameter. In fact the analysis has been restricted up to now only to small values of ϵ. It is easy to verify that $u = 0$ is a stationary solution which is stable for $\epsilon < 0$ and unstable for $\epsilon > 0$. This is very reminiscent of bifurcation theory. However we have here a bifurcation from a continuous spectrum which is a new situation. One observes that for positive ϵ there are two bands of unstable modes which are concentrated around the modes $k = \pm 1$ and have width $\mathcal{O}(1)\sqrt{\epsilon}$. To each mode of these bands is associated a periodic stationary solution of the SH equation. More precisely, we have the following result (see Collet-Eckmann 1987 for a proof).

Proposition 5. *Given a positive number K, there is a positive number ϵ_0 such that if $\epsilon \in]0, \epsilon_0]$, then for any number ω satisfying*

$$0 \leq (1-\omega^2)^2 \leq K\epsilon \,,$$

there is a periodic stationary solution S_ω of the SH equation of period ω. Moreover, modulo a constant translation, this solution is given by

$$S_\omega(x) = \epsilon^{1/2}\Gamma\cos(\omega x) + \mathcal{O}(\epsilon) \,,$$

where Γ satisfies

$$(1-\omega^2)^2 = \epsilon(1 - 3\Gamma/4) \,.$$

The main difference here with a bifurcation from a discrete spectrum is the occurrence of a one parameter family of solutions. As we can expect, this will produce a marginal mode in the stability analysis.

One should now determine the spectrum of the linearized evolution equation around one of the above stationary solution. This is more difficult than in the previous cases because we obtain an operator with periodic coefficients. Fortunately, one can use and control perturbation theory to get the following result.

Proposition 6. *Under the hypothesis of Proposition 5, if ω satisfies*

$$|\omega^2 - 1| \leq (\epsilon/3)^{1/2} \,,$$

the stationary solution S_ω constructed in Proposition 5 is marginally stable. If the above inequality is reversed, the solution is unstable (Eckhaus instability).

We refer to Collet-Eckmann 1987 for a detailed proof of this statement.

We now come to the construction of fronts for the SH equation between the unstable stationary solution $u = 0$ and one of the marginally stable solution S_ω given by Propositions 5 and 6. As we have already explained, we have to face a new difficulty which is that there is no frame of reference where the front will be stationary. This can be seen as follows. In the laboratory frame, the front is moving, but if we go to the moving frame, it is now the stationary part S_ω which becomes time dependent (in fact oscillating) because it is moving backward. This implies that we have to generalize our definition of fronts to incorporate the case of non spatially constant stationary solutions. This can be done as follows. We shall call a front of the SH equation from 0 to S_ω at speed c a function of two variables $W(.,.)$ such that

$$\begin{cases} W(x,-\infty) = S_\omega(x)\,, \\ W(x,+\infty) = 0\,, \\ W(x, x-ct) \text{ is a solution of the SH equation.} \end{cases}$$

One can of course think of more general extensions of the above definitions. In particular it should also be possible to incorporate phases which are time dependent corresponding to spatial intermittency for example. As we shall see below, the above definition is general enough for our present purposes.

We consider now the problem of fronts for the SH equation with a small and positive parameter ϵ. It is more convenient in this regime to use another small parameter $\eta = \epsilon^{1/2}$. It follows easily from dimensional considerations that one has to look for a front of the form $W(\eta x, \eta x - c\eta^2 t)$ in order to have finite limits for the speed c and the function W when $\eta \to 0$. Note that the speed is $c\eta$. With these notations we have the following result (Collet-Eckmann 1986).

Theorem 7. *Given a positive number K, there is a positive number ϵ_0 such that if $\epsilon \in]0, \epsilon_0]$, then for any real number ω satisfying*

$$(1-\omega^2)^2 \leq K\epsilon^2\,,$$

and for any number c near 2, there is a front from 0 to S_ω at speed $c\eta$.

The proof of the above theorem is rather lengthy. We shall only discuss here the main ideas. First of all, we look for a function W which is a superposition of Fourier modes in the first variable. In other words, we look for a moving envelope function for each fourier component of the stationary solution. It is these envelopes which have a constant shape in the frame moving at the speed of the front. Next one can start looking for a perturbation expansion for W in the small variable η. The main term is of course directly associated to the bifurcation. It is given by

$$e^{ix} A(\eta x, \eta^2 t) + \text{ c.c.}\,,$$

where A is a solution of the amplitude equation (CNW)

$$\dot{A} = A'' + cA' + A - A|A|^2\,.$$

The boundary conditions tell us that we should consider a solution of this equation which is a front. One can then continue the perturbation expansion at least formally. However one meets at this point two problems. The first one is that to the next order, the differential problem is degenerate, namely one gets a differential operator with a vanishing coefficient of the leading differential order if $\eta \to 0$. It is well known that such problems may have solutions with a singular dependence in the parameter η. A well known example is the semiclassical limit in quantum mechanics which is solved by a WKB approximation. Here, there are two types of solutions, one is well behaved and the other one is singular. Selecting at each level of perturbation theory in η the regular solution, one can construct a formal power series for a front. It is however not clear at this stage that the singular solutions do not occur when one wants to estimate the remainders. The second problem is that free integration constants appear at each level of perturbations. The two problems are in fact coupled. One can show that the remainder of the perturbation expansion can be controlled on the subset $(-\infty, (\log\eta)^2]$ of the real line. Outside this set, we expect the front solution to be very small, and it is then possible to use the theory of dynamical systems with an unstable fixed point. Without going into the details, we obtain a situation where only an infinite dimension (and infinite codimension) stable manifold is defined. The matching is then possible because there are enough (an infinite number) free integration constants to ensure that at $x = (\log\eta)^2$ the solution coming from $-\infty$ can hit the stable manifold. One verifies that after the matching there is only one free constant left which corresponds to a constant translation of the front. The fact that perturbation theory cannot be used everywhere can now be simply understood. The true front and the front of the CNW with the same speed do not have exactly the same decay rate at $+\infty$. This is however a crucial ingredient for the matching with the stable manifold.

The next step is to investigate the various stability questions required by the marginal stability criterion (Dee-Langer 1983). The spectral problems are however much more difficult to solve than before and we shall only investigate the essential spectrum. There is yet no result about an eventual non essential spectrum. The result can be formulated as follows (Collet-Eckmann 1987).

Theorem 8. *Under the hypothesis of Theorem 7, there is a positive constant K_* and a positive number ϵ_0 such that if $\epsilon \in]0, \epsilon_0]$, and ω satisfies*

$$(1-\omega^2)^2 \leq K_*\epsilon^2 ,$$

then Theorem 7 applies and the front from 0 to S_ω satisfies the marginal stability criteria (for the essential spectrum).

The converse of this result is true in the following sense. If the above inequality for ω is violated but the conditions of Theorem 7 are still satisfied, then the front does not satisfies the marginal stability criterion.

The proof of Theorem 8 is similar to the proof of Proposition 6. The difficulty comes from the fact that we have to find the spectrum of a differential operator with varying coefficients. This can be done using perturbation theory for the essential spectrum.

References

Aronson, D., Weinberger, H. (1978): Adv. Math. **30**, 30.

Bramson, M. (1983): Mem. Amer. Math. Soc. **285**.

Collet, P., Eckmann, J.-P. (1986): Commun. Math. Phys. **107**, 39.

Collet, P., Eckmann, J.-P. (1987): Helv. Phys. Acta **60**, 969.

Collet, P., Eckmann, J.-P. (1989): *Instabilities and fronts in extended systems*. Princeton University Press, Princeton (to appear).

Dee, G., Langer, J. (1983): Phys. Rev. Letter **50**, 383.

Landau, L., Lifshitz, E. (1959): *Statistical Physics*. Pergamon Press, London.

Sattinger, S. (1977): Journ. Diff. Equ. **25**, 130.

AN INTRODUCTION TO THERMOCAPILLARY CONVECTION

J.K. Platten and D. Villers

State University of Mons, 21, avenue Maistriau, B-7000 Mons.

1. Introduction

Natural convection is usually induced by density differences in the gravitational field, but may also be caused by a surface tension (or in interfacial tension) gradient that exists for some reasons along a liquid/gas or a liquid/liquid interface. This effect seems to be described for the first time by the italian physicist Carlo Marangoni [1] and therefore convection caused by surface tension effects has also been termed "Marangoni convection". This effect has also been described about the same time by the belgian Van der Mensbrugghe [2]. An historical introduction to these Marangoni effects may be found in ref. [3].

Marangoni convection is best described by a simple "kitchen experiment". Take a Petri dish, filled with water and let gently fall a drop of hexanol along the side of the dish; as soon as the alcohol has reached the surface, a strong motion develops from the low surface tension region (where the drop of alcohol has been placed) to the high surface tension region (pure water).

A few streamlines has been drawn on Fig. 1.

The recirculation is due to the boundaries. Due to the viscosity of the liquid this "surface convection" propagates into the bulk.

This type of convection is of technological importance. For example in the purification process of a crystal by the floating zone technique we have clearly a surface tension gradient along the free surface of the floating zone and therefore convective cells are observed inside the liquid bridge. In some circumstances, instead of to concentrate the impurities at the end of the rod, striations of impurities are observed with a rather well defined distance between the stripes. The striations are sometimes related to oscillatory convection. Therefore convection in the melt phase is related to the quality of the final product. As another example, in microgravity conditions (in space laboratories) convection cannot be related to density differences, but is caused by surface tension gradients. Thus crystal growth in space is another situation to which Marangoni convection could apply. Everybody knows that a heat or a mass flux is enhanced by convection; and in some instances one wants to have the highest possible fluxes; Marangoni convection is one of the possible convective transport mechanisms. Many other examples could be given.

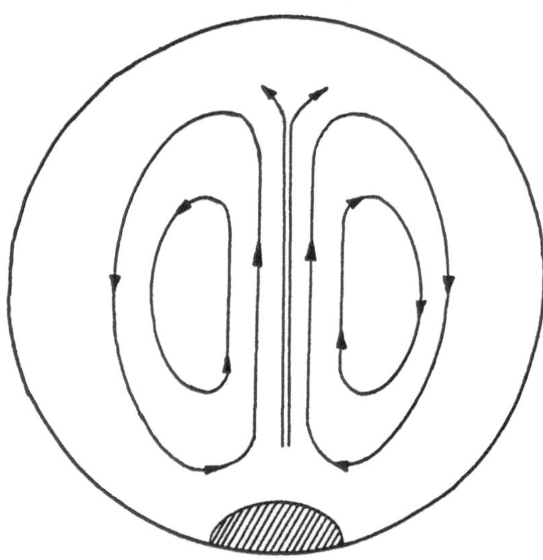

Fig. 1. Sketch of a simple experiment demonstrating the Marangoni effect

2 The experiments

There is thus a need for a "basic" (i.e. "academic") research on Marangoni convection in order to understand more precisely the phenomenon, to validate the models and to describe the different convective regimes. And when we say "basic research", that means that we need a simple geometry and a perfect control of the surface tension gradient $\partial\sigma/\partial x$. As a matter of fact, in our "kitchen experiment" the surface tension gradient due to a concentration gradient of some surface active substance, which is not so easily controled since it is clearly time dependent. It is much more easy to control a temperature gradient $\partial T/\partial x$ and since the surface tension is temperature dependent, one easy way to control a surface tension gradient is to impose a temperature gradient along an interface. The disadvantage is that the surface tension variation is rather small. Since for most liquids $\partial\sigma/\partial T \simeq -0.1 mNm^{-1}K^{-1}$, a temperature difference of say $10K$ will produce a surface tension variation of 1 or $2mNm^{-1}$, much smaller than the difference of surface tension between water and hexanol (say $50mNm^{-1}$). Yet the surface velocities are rather high, typically 5 times greater than the velocities observed in pure thermal convection when buoyancy acts alone and $\partial\sigma/\partial T = 0$.

We will present in this introductory paper a few results on convection; the system to be studied is sketched on Fig. 2.

The results we will show are experimental determinations of velocity profiles and comparison with theoretical predictions or numerical simulations. The experiments were performed using laser-Doppler velocimetry (LDV) which is known to be a non intrusive technique (there is no material probe in the liquid), highly precise and that needs no

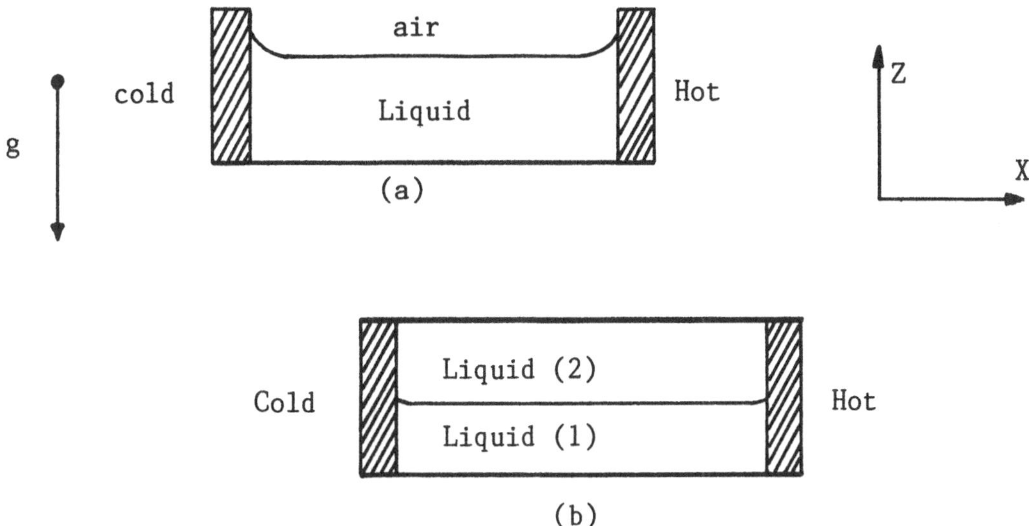

Fig. 2. Sketch of a simple situation discussed in this paper

calibration. However the method requires a transparent fluid. The description of the equipment we have can be found elsewhere [4]. Fig. 3 shows a measured horizontal velocity profile in a layer of acetone 2.5 mm thick (the other dimensions of the rectangular cavity are 30 mm and 10 mm). The horizontal velocity profile of Fig. 3 was taken along the vertical median; the conditions were

$$T_{Hot} - T_{Cold} = 1.2K$$

at a mean temperature of $21°C$.

It may be observed that the surface velocity is much greater than anywhere in the bulk. Regarding the theoretical predictions, the system is described by the conservation equations of macroscopic physics : the momentum balance equation (or the so-called Navier-Stokes equations) and the energy equations, together with appropriate boundary conditions. These equations are nonlinear partial differential equations and need a numerical treatment. They have been solved by a finite differences technique in the two-dimensional approximation [5]. Using a 193 × 33 grid (i.e. 6369 grid points) we have to solve in the streamfunction-vorticity-temperature formulation a system of 19107 ordinary differential equations. This has been done for the real experiment presented on Fig. 3 (i.e. with the correct values for all the physical properties or transport coefficients such as thermal expansion coefficient, viscosity, density, thermal conductivity, surface tension variation with temperature, etc..). The computed velocity profile is the full line on Fig. 3. The agreement between the laboratory and the computer experiment is satisfactory : the

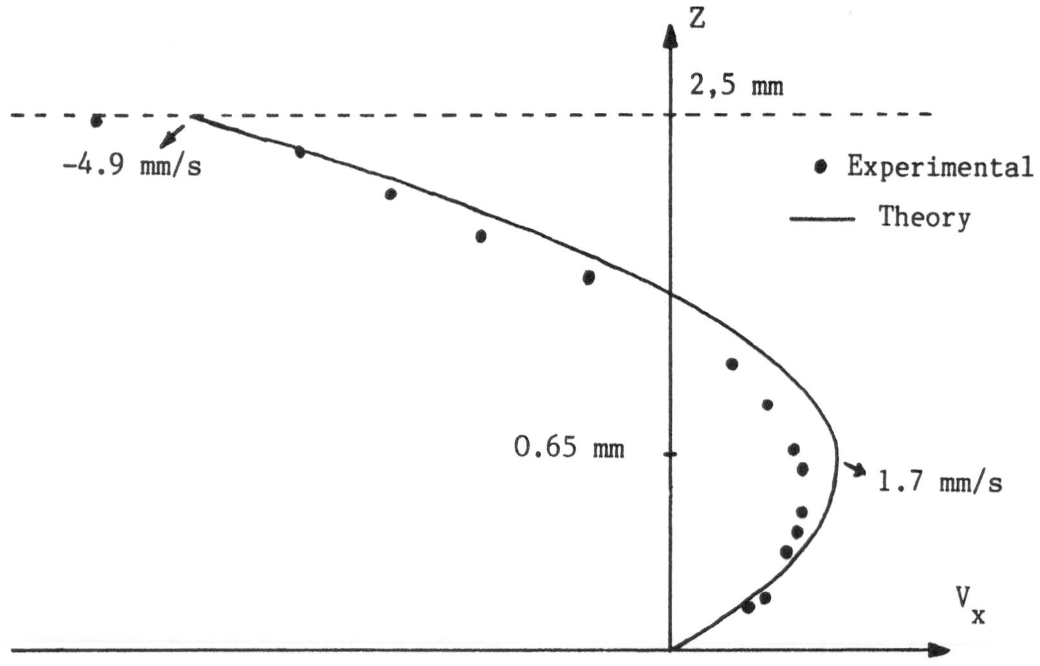

Fig. 3. Horizontal velocity profile on the vertical median of a cavity filled with acetone and submitted to an horizontal temperature gradient. Experiment and numerical simulation.

computed surface velocity is 4.9 mm s^{-1}, the observed one is 5.9 mm s^{-1}. By increasing the constraints (i.e. the temperature gradient and the depth of the layer) oscillatory convection has been observed in acetone (for Δ T = 6K; layer 5.9 mm thick) with a rather small period of oscillations ($\simeq$ 13 seconds). A comparison with theoretical predictions is now undertaken but the first results seem to indicate also a correct order of magnitude. The nature of the new oscillatory state is not easy to understand. One could imagine a kind of traveling wave solution with, at a given point, a strictly sinusoidal time variation of the velocity. This is what is predicted by a linear hydrodynamic stability theory, done for a lateraly infinite system and has also been observed by numerical simulations using lateral periodic boundary conditions in order to simulate an infinite system. But as soon as lateral rigid boundaries are introduced, with no-slip boundary conditions (zero velocity along such a boundary), the scenario is quite different. It seems that the unique basic convective cell is splitted into many small cells of unequal size but the state of the system remains steady. One has to increase further the constraints in order to find oscillatory convection, which is not of the traveling wave kind. Rather, it seems that the cell sizes, the cell positions and even the number of cells oscillate in time. One of the unresolved question is then : can we extrapolate results for infinite systems to finite systems, and how big has the aspect ratio to be in order to learn something from infinite system. And if we cannot extrapolate, then the only way is probably numerical simulation of the flow field. And even if we do such simulations, can we learn something from two-dimensional simulation to apply it to the real three dimensional world? Nevertheless one has to carry out research on Marangoni convection in order to help engineers working in material processing (on earth or in microgravity) where convection (oscillatory or steady) can play an important role. The case of thermal convection in superposed immiscible liquid layer (Fig 2b) raises some fundamental question in fluid dynamics : suppose that in both layers hot fluid raises and cold fluid sinks, giving rise to an anticlockwise circulation cell in each layer (Fig. 4)

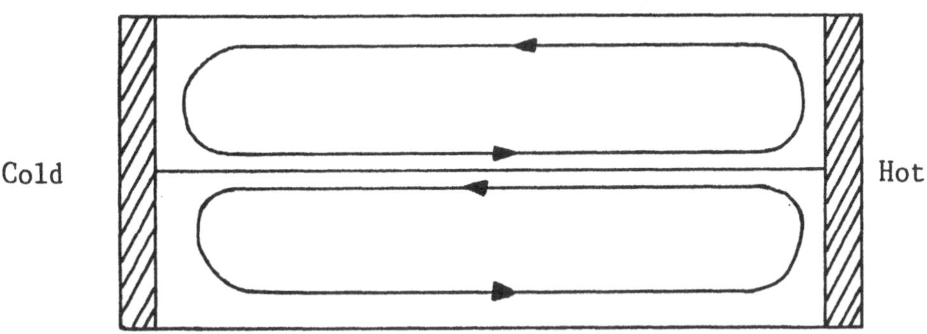

Fig. 4. Possible convective celles in two immiscible liquid layers.

Clearly the interface should be at rest ($V_x = 0$) which is not a required condition: physics asks only the continuity of the horizontal velocity at the interface. Thus the fundamental question is to know the kind of coupling between the two cells. If thermal effects are dominant the circulation shown on Fig. 4 should prevail. On the contrary, "mechanical coupling" should produce a third contrarotative cell as shown for example on Fig. 5; but the question is to know in which layer is located this intermediate cell which accomodates the shear stress at the interface.

We have done an experiment with a 3.7 mm thick heptanol layer on top of a 5.6 mm thick water layer (with a T = 12.8 K on the 30 mm interface) and the recorded velocity profile is given on Fig. 6a

The observed velocity profile is compatible with a pattern of convection shown on Fig.6b. The most striking result is an acceleration of the interface towards the hot boundary and this will be explained later. One could think (but this is very naïve) that the intermediate convective cell with clockwise rotation is located in the water layer because of its higher thickness. Geometrical considerations are not sufficient to explain the observed facts. We have now a model for this convection which takes into account not the ratio of the thicknesses of the two layers but also the ratio of the expansion coefficients and of the viscosities. Interfacial tension gradients are also taken int account [6]. Indeed if the momentum balance equation at the interface is incorrectly written as

$$\mu^{(1)} \frac{\partial V_x^{(1)}}{\partial x} = \mu^{(2)} \frac{\partial V_x^{(2)}}{\partial z}$$

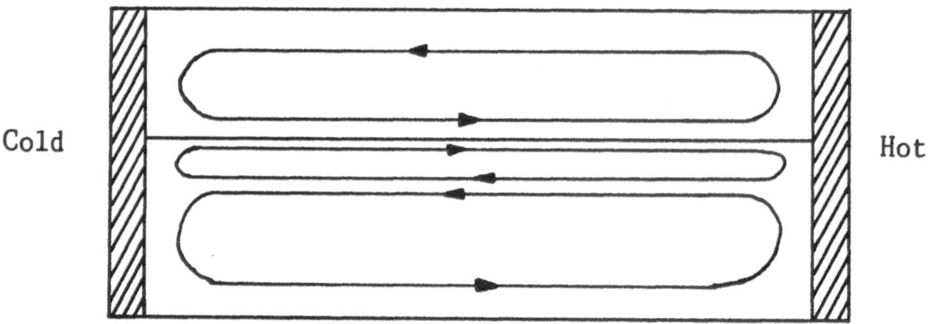

Fig. 5. Possible convective cells with mechanical coupling in two immiscible liquid layers.

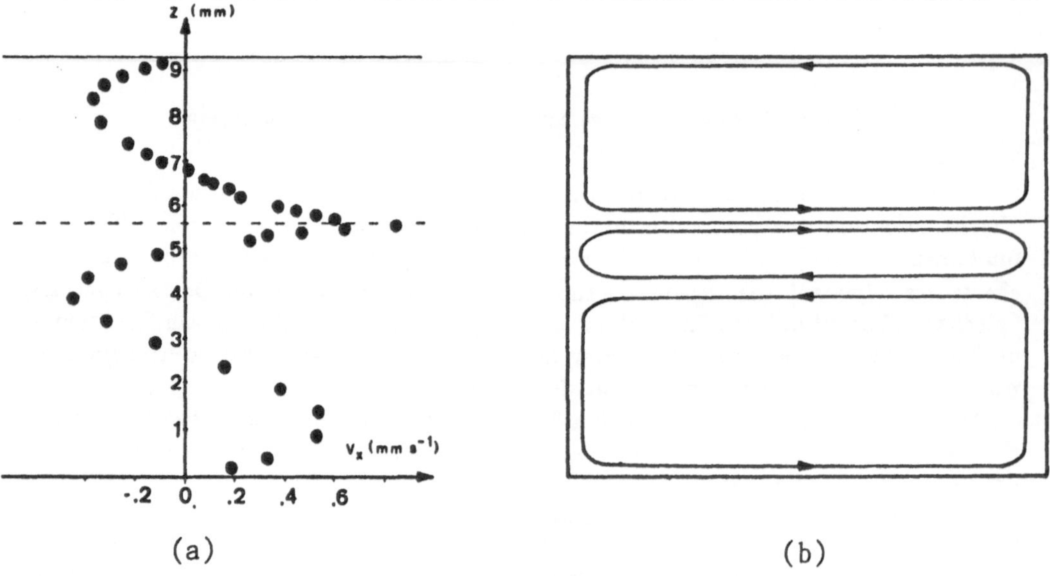

Fig. 6. Horizontal velocity profile (a) and compatible convective cells (b) for a layer of 3.7 mm heptanol above a 5.6 mm water layer.

(continuity of the tangential stress), there cannot be a change of sign in the slopes of the velocity gradients near the interface as observed really on Fig. 6a. Therefore following Levich (see e.g. [7]) we write more correctly

$$\mu^{(1)}\frac{\partial V_x^{(1)}}{\partial z} - \mu^{(2)}\frac{\partial V_x^{(2)}}{\partial z} = \frac{\partial \sigma}{\partial x} \simeq \frac{\partial \sigma}{\partial T}.\frac{\partial T}{\partial x}$$

From a few points in each layer near the interface we may estimate the velocity gradient (e.g. $\partial V_x^{(1)}/\partial z \simeq 1.325 s^{-1}$ and $\partial V_x^{(2)}/\partial z \simeq -0.947 s^{-1}$); the viscosities of each layer are also known ($\mu^{(1)} \simeq 1.110^{-2}$ poise; $\mu^{(2)} \simeq 7.10^{-2}$ poise); an estimation of the temperature gradient is 12.8 K/ 30 mm. Therefore, as a byproduct of the LDV experiment and the determination of the velocity profiles, we get an estimation of $\partial\sigma/\partial T$.

$$\frac{\partial\sigma}{\partial T} = +0.018 mNm^{-1}K^{-1}$$

It is indeed very strange and unusual that interfacial tension increases with temperature but on the other hand this explains the acceleration of the interface from cold wall to hot wall, i.e. from low surface tension to high surface tension regions.

Immediately we have undertaken a direct measurement of the surface tension in function of the temperature using the Wilhelmy plate method and the Tensiomat 2000 from Prolabo (resolution $\simeq$ 0.05 mN m^{-1}). The experiments were done between -3° and 80°C and the result is given on Fig.7.

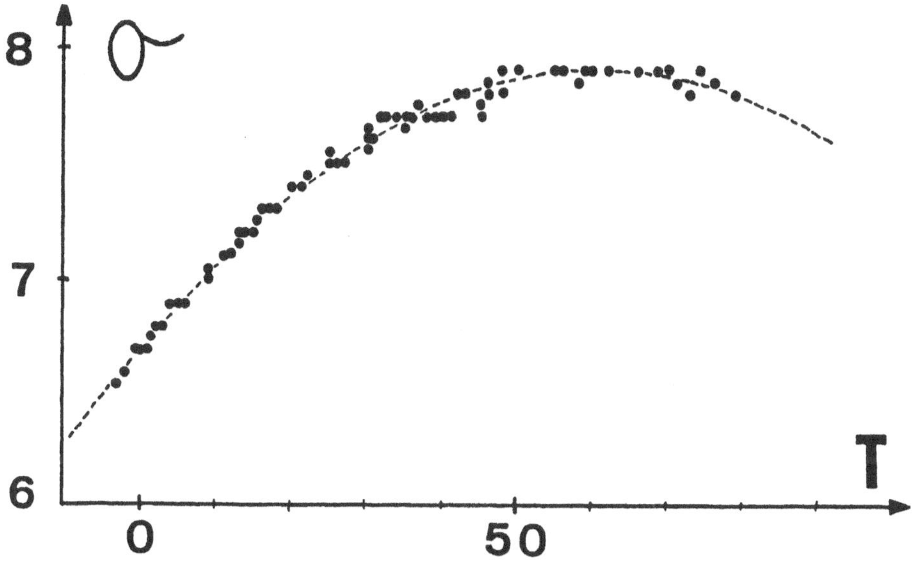

Fig. 7. Temperature dependence of the interfacial tension between water and *n*-heptanol (σ in mN m^{-1} and T in °C).

At 21.3° (the temperature at which the LDV experiment was performed), we find

$$\left[\frac{\partial\sigma}{\partial T}\right]_{21.3°C} \simeq +0.022 mNm^{-1}K^{-1}$$

i.e. a value close to the indirect determination. Fig. 7 shows a maximum in the curve $\sigma(T)$. By changing the nature of the alcohol [8] (from butanol to dodecanol) the same behaviour was always observed : a parabolic variation of the surface tension with temperature. The temperature at which the maximum occurs is an increasing function of the number of carbon atoms in the chain (ranging from 20°C for butanol to 89°C for dodecanol). A proposition of an explanation for such extrema has now been undertaken, using statistical mechanic theory [9]. The corresponding interfacial tension between water and the alkanes shows a standard linear decrease with temperature : for example σ_{wa} has a temperature derivative $\partial\sigma/\partial T = -0.07 mNm^{-1}K^{-1}$ constant between 25°C and 80°C.

3 Concluding remarks

In conclusion, interfacial or surface tension gradients (due to thermal or concentration gradients) induce surface convection which propagates in the bulk. Studying experimentally by a LDV technique the convective pattern, we have discovered that interfacial tension between water and n-alcohols always increases with temperature in some range (parabolic behavior) and this remains to be explained. On the other hand Marangoni convection is of large technological importance : in hydrometallurgy (crystal growth from the melt; crystal purification...), in thermal methods in oil recovery, in liquid-liquid extraction were convection may enhance the transfer of a solute through an interface. the examples we presented above, Marangoni convection is of "macro-scale" type, i.e. produced by a macro-scale gradient. "Micro-scale" gradient, produced by small temperature or concentration inhomogeneities, can grow in time if there is a particular "Fourier mode" which is unstable for the given set of external parameters and this particular mode saturates, giving rise to macroscopic observable motion. As a last example of such a situation, let us consider [10] an oil lens (of nitroethane) on water containing a surfactant.

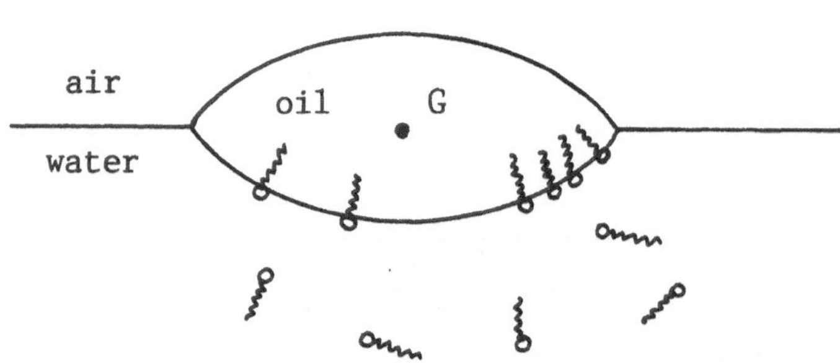

Fig. 8. Oil lens and transfer of solute through an interface.

Equilibrium can be reached by a purely diffusive transfer and an adsorption-desorption process at the interface. Typically for a pure diffusive transfer the time needed is hour. But small inhomogeneities of the surfactant at the liquid/liquid interface (a so-called fluctuation) may give rise to an hydrodynamical instability : the center of gravity G of the lens starts to oscillate, sometimes superposed to a general drift; also the transfer of solute is enhanced by convection, equilibrium is reached only after 2 minutes! We hope to have convinced the reader of the importance of thermal or solutal Marangoni convection.

References

1 Marangoni, C.G.M., Ann. Phys. 143, p337, 1871.

2 Van der Mensbrugghe, G.L., Mém. courronnés Acad. Roy. Belg., 34, 1869.

3 Scriven, L.E., Sterling, C.V.: "The Marangoni effects", Nature, 187, p186-188, 19

4 Platten, J.K., Villers, D., Lhost, O.: "DV study of some free convection problem extremely slow velocities: Soret driven convection and Marangoni convection", in laser anemometry in fluid mechanics III, p245-260. Editors: Adrian, Asanuma, Durao, Durst and Whitelaw, Ladoan, Lisbon, 1988.

5 Villers, D., Platten, J.K.: "Marangoni convection in systems presenting a minimum surface tension", PhysicoChemical Hydrodynamics, 6(4) (1985) pp. 435-451. and Villers, D., Platten, J.K.: "Influence of thermocapillarity on the oscillatory convection in low Pr fluids.", Proceedings of the GAMM Workshop on "Numerical simulation of oscillatory convection in low Pr fluids", Marseille, France, October 1988 (to appear 1989).

6 Villers, D., Platten, J.K.: "Influence of interfacial tension gradients on thermal convection in two superposed immiscible liquid layers". Applied Scientific Research, to appear.

7 Landau, L.D., Lifshitz, E.M.: Fluid Mechanics, chapter VII, p234, Pergamon press (1959).

8 Villers, D., Platten, J.K.: "Temperature dependence of the interfacial tension between water and long-chain alcohols", J. Physical Chemistry, 92p 4023-24, 1988.

9 De Coninck, J., Villers, D., Platten, J.K.: "Non-monotonous temperature dependence of interfacial tensions", in preparation

10 Bekki, S.: "Contribution à l'étude de l'effet Marangoni de soluté", thèse, université de Paris VI, 1988.

ANTONOV RULE AND MULTILAYER WETTING IN "q STATE" MODELS

Alain Messager

Centre de Physique Théorique
CNRS-Luminy, Case 907
F-13288, Marseille Cedex 9 (FRANCE)

1. INTRODUCTION

In this lecture we shall be interested in the wetting phenomena in the grand canonical ensemble , namely we shall study the validity of the Antonov rule, which was described by Gibbs [1]:

Let A, B, C, be three different fluid phases of matter, which satisfy the conditions necessary for equilibrium when they meet at plane surfaces. The components of A and B may be the same or different, but C must have no components except such as belong to A or B. Let us suppose masses of the phase A and B to be separated by a very thin sheet of the phase C [...]. The value of the superficial tension for such a film will be $\sigma_{AC} + \sigma_{BC}$ *, if we denote by these symbols the tensions of the surfaces of contact of the phases A and C, and B and C, respectively [...]. This value will not be affected by diminishing the thickness of the film, until the limit is reached at which the interior of the film ceases to have the properties of matter in mass. Now if* $\sigma_{AC} + \sigma_{BC}$ *is greater than* σ_{AB} *, the tension of the ordinary surface between A and B, such a film will be at least practically unstable. We cannot suppose that* $\sigma_{AB} > \sigma_{AC} + \sigma_{BC}$ *, for this would make the ordinary surface between A and B unstable and difficult to realize. If* $\sigma_{AB} = \sigma_{AC} + \sigma_{BC}$ *, we may assume, in general, that this relation is not accidental, and that the ordinary surface for the contact for A and B is of the kind that we have described .*

The problem can be set up analogously when several phases (more than three) $A_1,A_2,....,A_n$ coexist. An interface between two of them say A_i, A_{i+2}, may be wetted by a layer of a third one A_{i+1}. The condition of perfect wetting of A_i, A_{i+2}, by a film of A_{i+1} can be expressed in terms of the *spreading coefficient:*

$$s(A_i,A_{i+1},A_{i+2}) = \sigma(A_i,A_{i+2}) - \sigma(A_i,A_{i+1}) - \sigma(A_{i+1},A_{i+2}).$$

The interface between the phases A_i, A_{i+2} should be wetted by the phase A_{i+1} when the *Antonov rule* is satisfied: $s(A_i,A_{i+1},A_{i+2}) = 0$. In particular in the case of three coexisting phases if $s(A_1,A_2,A_3) = 0$, we will obtain the *single layer wetting.* More generally if all the phases coexist and all the spreading coefficients $s(A_i,A_j,A_k)$, with $i< j< k$, vanish the interface between the phases A_1 and A_n will be wetted by the sequence of films of the phases $A_2,...,A_{n-1}$. This is *multilayer wetting.* For this purpose we introduce a new class of "q state" models which can exhibit, when the temperature increases, phase transitions between phases of decreasing order ; if we choose conveniently the couplings all these phases can coexist at some temperature ; then multilayer can occur, this will be the content of the second part of this lecture.

We first study *perfect wetting* in the Potts model.

2 PERFECT WETTING IN THE POTTS MODEL

The "q state" Potts model is well known to have a transition temperature T_t where q ordered phases coexists with a disordered phase [2],[3] for q large enough and for any dimension $d\geq2$. Two kinds of interfaces may therefore appear within this context: *one between the ordered phases and one between the ordered and the disordered phases.*

The Potts Hamiltonian in a finite box Λ for a configuration $x_\Lambda \equiv \{x_i\}$ $i\in\Lambda$ with boundary condition $\tilde{x}$ (i.e. a configuration on $\mathbb{Z}^d$) is:

$$H_\Lambda^{b.c.}(x_\Lambda \mid \tilde{x}) = - \sum_{<i;j>\cap\Lambda\neq\emptyset} \delta(x_i,x_j) - \sum_{i\in\Lambda, j\in\Lambda^c} \delta(x_i,\tilde{x}_j)$$

where $x_i \in \{1,...,q\}$, $q\geq2$, $J\geq0$ and δ is the usual Kronecker symbol. The summation is restricted over nearest neighbors.

We denote by $\langle\ \rangle_\Lambda^{b.c.}$ the expectation corresponding to the Gibbs measure μ_Λ:

$$\mu_\Lambda = [Z_\Lambda^{b.c.}(\beta)]^{-1} \exp\{-\beta H_\Lambda^{b.c.}\}.$$

In the following, we shall study the following type of boundary conditions:

- (a) : the ordered b.c. : $\tilde{x}_i = a$, $a = 1,2,\dots q$, for every i in $\mathbf{Z}^d$
- (f) : the free b.c. where the sum runs only over the pairs $<i;j>$ included in Λ.
- (a,b): the mixed b.c.: $\tilde{x}_i = a$ if $z>0$; $\tilde{x}_i = b$ if $z<0$
- (a,f) b.c. is defined similarly by : $\tilde{x}_i = a$ above the plane and free b.c. below.

We are now able to define the surface tensions which appear in our problem. We consider a rectangular box Λ centred at the origin. Let S_Λ be the area of the portion inside Λ of the plane defined by z=0. We define the surface tension at the inverse temperature β by:

$$\sigma_\beta^{a.b} = \lim_{\Lambda\uparrow Z^d} -\frac{1}{S_\Lambda} \log \frac{Z_\Lambda^{a.b.}(\beta)}{[Z_\Lambda^a(\beta)\, Z_\Lambda^b(\beta)]^{1/2}}$$

$\sigma_\beta^{a.f}$ is defined analogously

Remark: we have considered all the possible surface tensions between the coexisting phases due to the symmetries of the model $\sigma_\beta^{a.b}$ has the same value for every choice of a and b.provided $a\neq b$ and $\sigma_\beta^{a.f}$ is independant of the choice of a.
The following properties of these surface tensions were proven in [4],[5]

Theorem 1 *For the q-states Potts model for q large, and $d \geq 2$, we have :*

$$\sigma_{\beta_t}^{a.b} > 0$$

$$\sigma_\beta^{a.b} > 0 \quad \text{for } \beta \geq \beta_t$$

The Gibbs Antonov rule can be proved for the Potts model at the transition temperature T_t.

Theorem 2. *For the q-states Potts model with q even and $d \geq 2$, we have at βt, and for any two different ordered phases a and b.*

$$\sigma^{a.b}_{\beta t} = \sigma^{a.f}_{\beta t} + \sigma^{b.f}_{\beta t} > 0$$

In particular, in dimension d =2:

$$\sigma^{b.f}_{\beta} = 0 \text{ for } \beta \neq \beta_t$$

We just give an idea of the proof: the first inequality is derived, for every β, from the following correlation inequality proved in [6]: for any subsets A and B of L, and $a \neq b$.

$$\langle \prod_{i \in A} \delta(x_i,a) \prod_{i \in B} \delta(x_i,b) \rangle - \langle \prod_{i \in A} \delta(x_i,a) \rangle \langle \prod_{i \in B} \delta(x_i,b) \rangle \leq 0$$

The second inequality in theorem 2 is obtained, at β_t only, by restricting the summation over a conveniently chosen set of interfaces [7].

The physical meaning of the Antonov rule is that there is a sheet of disordered phase which separates the two ordered phases a and b at the transition temperature. This is perfect wetting.

Remark: it is simple to realize that two ordered phases cannot be wetted by a third ordered phase because the Antonov rule cannot be fulfilled in this case.

3 DEFINITION OF A CLASS OF "q STATE MODELS".

Now we introduce a class of "q state" models [8] which are a generalization of the Potts model. They are defined by the following Hamiltonians:

$$H = - \sum_{<i,j>} \sum_{r=1}^{m} Jr \prod_{\alpha=1}^{r} \delta(x_i^{\alpha}, x_j^{\alpha})$$

where $x_i^{\alpha} \in \{1,....,q_{\alpha}\}$, $Jr \geq 0$, $\prod_{\alpha=1}^{m} q_{\alpha} = q$, so that

$$\sigma = x_i^1 + (x_i^2-1)q1 + (x_i^3-1)q1q2 + ... + (x_i^m-1)q1...qm-1$$

takes the values 1, ...q.

What makes these models interesting is that they could exhibit a cascade of phase transitions. At low temperature the pure phases should be totally ordered, then at some transition temperature a subgroup of Z(q) is broken and we have a new set of partially ordered pure phases; this can continues when the temperature increases with less and less ordered pure phases untill we obtain the disordered pure phase. For appropriate values of the interactions Jr all these pure phases could coexist.

Next, we introduce residual free energies for this class of models which under appropriate conditions (coexistence of the corresponding pure phases) will have the physical meaning of surface tensions or interfacial free energies. For this purpose we consider translation invariant and finite range interaction potentials and take the set Λ as a rectangular box centred at the origin on a d dimensional lattice $\mathbf{Z}^d$, and define, for $r = 0,1,\dots,m$ the following partition functions with boundary conditions:

$$Z_\Lambda(D_a^r) = \sum_{\sigma_\Lambda;\sigma_{\partial\Lambda}} \exp\{-\beta\, H(\sigma_\Lambda \mid \sigma_{\partial\Lambda})\} \prod_{i\in\partial\Lambda} \left(\frac{p_r}{q}\right) \delta_{p_r}(\sigma_i,a) .$$

$\partial\Lambda$ denotes the boundary of Λ (with a thickness equal to the range of the interaction)

$$p_r = q_1 \dots q_r$$

$\delta_p(\sigma,\sigma')=1$ if $\sigma=\sigma'$ modulo p and 0 otherwise and $H(\sigma_\Lambda \mid \sigma_{\partial\Lambda}) = H(\sigma_{\Lambda\cup\partial\Lambda}) - H(\sigma_{\partial\Lambda})$.
We introduce also mixed boundary conditions $(D_a^r, D_b^{r'})$ with respect to the plane z= 0 and the corresponding partition functions

$$Z_\Lambda(D_a^r \mid D_b^{r'}) = \sum_{\sigma_\Lambda;\sigma_{\partial\Lambda}} \exp\{-\beta\, H(\sigma_\Lambda \mid \sigma_{\partial\Lambda})\} \prod_{i\in\partial\Lambda^+} \left(\frac{p_r}{q}\right) \delta_{p_r}(\sigma_i,a) \prod_{i\in\partial\Lambda^-} \left(\frac{p_{r'}}{q}\right) \delta_{p_{r'}}(\sigma_i,b)$$

where $\partial\Lambda^+$ is the part of $\partial\Lambda$ located above the plane and $\partial\Lambda^- = \partial\Lambda \setminus \partial\Lambda^+$. We define the surface tension at inverse temperature β by:

$$\sigma(D_a^r, D_b^{r'}) = \lim_{\Lambda\uparrow \mathbf{Z}^d} -\frac{1}{\beta S_\Lambda} \log \frac{Z_\Lambda(D_a^r, D_b^{r'})}{[Z_\Lambda(D_a^r)\, Z_\Lambda(D_b^{r'})]^{1/2}}$$

Let us first study the above Hamiltonian for m=2. This model is a generalization of the *Ashkin Teller* model.

4 A GENERALIZED ASHKIN TELLER MODEL: 3 LAYER WETTING.

$$H(\sigma_\Lambda) = - \sum_{<i,j>\subset\Lambda} J_1\, \delta(x_i^1,x_j^1) + J_2\, \delta(x_i^1,x_j^1)\, \delta(x_i^2,x_j^2)$$
$$= - \sum_{<i,j>\subset\Lambda} J_1\, \delta_{q_1}(\sigma_i,\sigma_j) + J_2\, \delta(\sigma_i,\sigma_j)\,.$$

For this model we expect the following phase diagram:

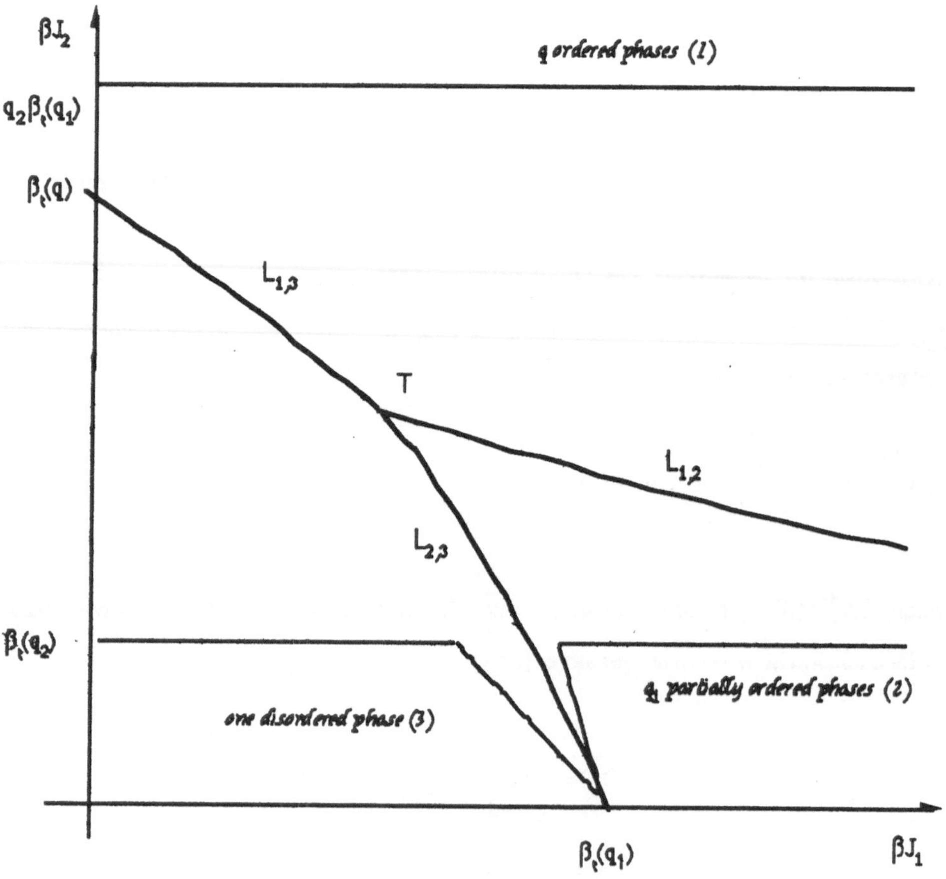

The solid lines correspond to the expected phase transition lines. The dashed lines bound the regions where the number and nature of the phases can be established rigorously. Let $a=a_1+ a_2(q_1-1)$, $a_i\in \{1,...,q_i\}$, we denote $\langle\,.\,\rangle_\Lambda^a(J_1,J_2)$ the expectation with respect to the conditional Gibbsian measure with closed "a" boundary condition:

$$\langle\,.\,\rangle^{a}_{\Lambda}(J_1,J_2)=Z^{-1}_{\Lambda}(a)\sum_{\sigma_\Lambda;\sigma_{\partial\Lambda}}\exp\{-\beta\, H(\sigma_\Lambda\,|\,\sigma_{\partial\Lambda})\}\prod_{i\in\partial\Lambda}\delta(\sigma_i,a)$$

and $\langle\,.\,\rangle^{a}(J_1,J_2)$ the infinite volume limit and let $\beta_t(q)\approx 1/d\ \ln q$ denote the transition point of the q-state d-dimensional Potts model. Let us introduce the two following magnetizations:

$$m_1=\langle\, q_1\ \delta(x^1_i,a_1)-1\rangle^{a}(J_1,J_2)$$

$$m_2=\langle\, q_2\ \delta(x^2_i,a_2)-1\rangle^{a}(J_1,J_2)$$

The following proposition is proved with correlation inequalities, as in ref.9. It gives a partial decomposition of the phase diagram.

Proposition 1. For q_1 and q_2 large enough we have:

a) If $\beta(J_1+J_2)<\beta_t(q_1)$ and $\beta J_2<\beta_t(q_2)$ then: $m_1=0$; $m_2=0$

b) If $\beta(J_1+\frac{J_2}{q_2})\geq\beta_t(q_1)$ and $\beta J_2<\beta_t(q_2)$ then: $m_1>0$; $m_2=0$

c) If $\beta(J_1+\frac{J_2}{q_2})\geq\beta_t(q_1)$ and $\beta\frac{J_2}{q_1}\geq\beta_t(q_2)$ then: $m_1>0$; $m_2>0$

The idea of a complete proof (which should use the Pirogov Sinai theory) of the phase diagram described in the above picture is based on the following remarks: Let us defined the following restricted ensembles:

-1 The q ordered ground states where all the spins are identical, whose weights are proportional to $\exp\{\beta(J_1+J_2)\, l(\Lambda)\}$, (where $l(\Lambda)$ is the number of n.n. pairs in Λ).

-2 The partially ordered restricted ensembles where $\sigma_i\in D^1_a$ and for all nearest neighbours i, j, $\sigma_i\neq\sigma_j \bmod q$, i.e. ($x^1_i=x^1_j$, $x^2_i\neq x^2_j$). When a=1,...,q there are q_2 different restricted ensembles D^1_a whose weights are proportional to $\exp\{\beta J_1\, l(\Lambda)+|\Lambda|\,[\ln q_2+O(1/q_2)]\}$

-3 The disordered restricted ensemble denoted D where all spins are different, whose weight is proportional to $\exp\{|\Lambda|\,[\ln q+O(1/q)]\}$

At the point $P=(\beta J_1\approx 1/d\ \ln q_1,\beta J_2\approx 1/d\ \ln q_2)$ the three weights are equal: all the phases should coexist. There are three lines starting from P where the weights 1 and 2 (resp. 1 and 3),(resp.

2 and 3) are equal, on which the associated phases should coexist. There are three regions 1, 2, 3, where the corresponding weight is dominant.

Remark: if we choose J_1/J_2 large enough we see the occurrence of an intermediate phase

Remark: it is interesting to compare this phase diagram with that of the Ashkin-Teller model see for example ref.[10] p.362 or of the Abelian Higgs model.

We prove the posivity of the "spreading coefficient" associated to the different surface tensions in the following proposition using the correlation inequalities of the appendix.

Proposition 2.

$$\sigma(a,b) \geq \sigma(a,D) + \sigma(D,b)$$

$$\sigma(D_a^1,D_b^1) \geq \sigma(D_a^1,D) + \sigma(D,D_b^1)$$

$$\sigma(a,b) \geq \sigma(a,D_a^1) + \sigma(D_a^1,D_b^1) + \sigma(D_b^1,b)$$

$$\sigma(a,b) \geq \sigma(a,D_a^1) + \sigma(D_a^1,D) + \sigma(D,D_b^1) + \sigma(D_b^1,b)$$

Then according to the above discussion , we expect that :

-The disordered phase wets the interface between the ordered phases on the line L13, and wets the interface between the partially ordered phases on the line L23.

-The partially ordered phases wet the interface between the ordered phases on the line L12.

-At the triple point T there is *3 layer wetting.* This situation is described in the following picture :

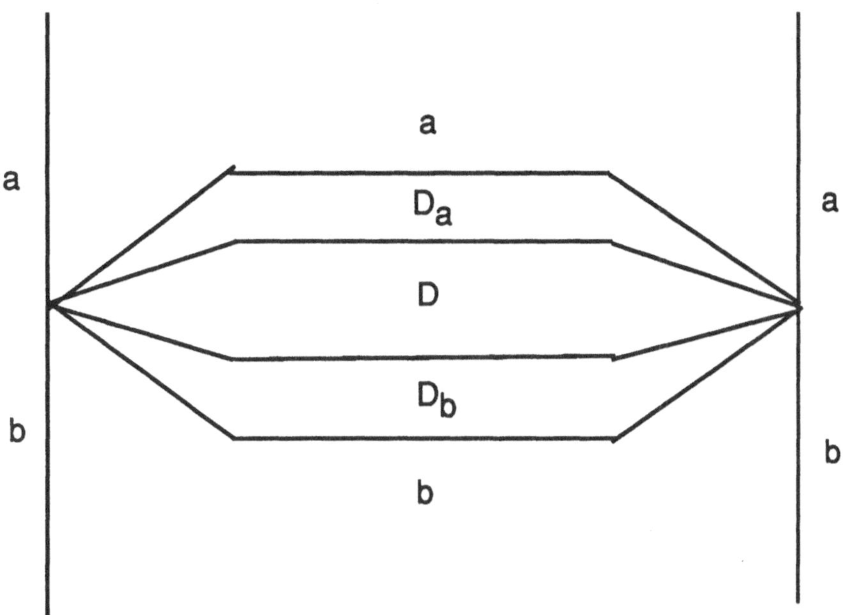

Warning :We have no macroscopic contact angle.

5 The "q STATE" MODEL: MULTILAYER WETTING

We finally consider the most general Hamiltonian of section 3: let us define the following restricted ensembles:

-The $q=p_m$ ordered configurations, a=1,...,q, whose weight are proportional to $\exp\{\beta l(\Lambda)\sum_r J_r\}$

-The partially ordered restricted ensembles of type r ($1 \le r \le m-1$) where $\sigma_i \in D_a^r$ and for all n. n. pairs i, j, $\sigma_i \neq \sigma_j \bmod p_{r+1}$ i.e. ($x_i^\alpha \neq x_j^\alpha$ for $\alpha \ge r+1$, $x_i^\alpha = x_j^\alpha$ if $\alpha \le r$).
When a=1,...,q, there are p_r different restricted ensembles D_a^r, whose weights are proportional to $\exp\{\beta l(\Lambda)\sum_1^k J_k + |\Lambda| [\ln q/p_r + O(p_r/q)]\}$

-The disordered restricted ensemble where for all n. n. pairs i, j, $\sigma_i \neq \sigma_j$, i.e. ($x_i^\alpha \neq x_j^\alpha$ for $1 \le \alpha \le m$) whose weight is proportional to $\exp\{|\Lambda| [\ln q + O(1/q)]\}$.

For large p_r, all these weights are equal at some point $\beta J_r \approx 1/d \ln p_r$ and according to the previous discussion, we expect that there is a unique point $\beta J_r \approx 1/d \ln p_r$ where all the corresponding phases coexist

The following propositions are proved with the correlation inequalities stated in the Appendix.

Proposition 5. *If $r \geq r''$ and $r' \geq r''$, $a \neq b$ (mod p_1), we have:*

$$\sigma(D_a^r, D_b^{r'}) \geq \sigma(D_a^r, D_b^{r''}) + \sigma(D_a^{r''}, D_b^{r''}) + \sigma(D_a^{r''}, D_b^{r'})$$

Proposition6.

$$\sigma(a,b) \geq \sigma(a, D_a^{m-1}) + \sigma(D_a^{m-1}, D_a^{m-2}) + .. + \sigma(D_a^1, D) + \sigma(D, D_b^1) + ... \sigma(D_b^{m-2}, D_b^{m-1}) + \sigma(D_b^{m-1}, b)$$

According to the Gibbs argument these propositions lead to *2m-1 layer wetting.*

CONCLUSION:

We believe that all the inequalities between surface tensions, in the multilayer case, can be converted into equalities when the corresponding phases coexist, as it has been proved for the Potts model. Another important question is related to the thickness of the film either in the case of perfect wetting or in the case of multilayer wetting.

ACKNOWLEDGMENTS

I want to thank F. Dunlop and J. De Coninck and for their kind invitation. This talk is essentially based on joint works with J. De Coninck, F. Dunlop, R. Kotecky, L.Laanait S. Miracle Sole, J.Ruiz, and S.B. Shlosman. Thanks to them.

APPENDIX : Correlation Inequalities.

Let Λ be a finite set of sites. To each $i\in\Lambda$ we attach a spin variable σ_i which may assume q values $\{1,2,...,q\}$ with $q \geq 2$. The energy of a configuration $\sigma_\Lambda \in \{1,....,q\}^{|\Lambda|}$ is given by:

$$H(\sigma_\Lambda) = - \sum_{\{i;j\}\subset\Lambda} J_{ij}\, \delta(\sigma_i,\sigma_j) - \sum_{i\in\Lambda} h_i(\sigma_i)$$

We assume that $J_{ij} \geq 0$ for all $\{i,j\}\subset\Lambda$, and that the functions h_i are such that, for some integer k satisfying $1\leq k \leq q-1$ and for all $i\in\Lambda$, we have

$$\sum_{a=1}^{k} h_i(a) \cos \frac{2\pi na}{k} \geq 0\,, \quad h_i(n) = h_i(k-n)\,, \quad \text{for every } n=1,...,k-1,$$

and

$$\sum_{a=k+1}^{q} h_i(a) \cos \frac{2\pi n(a-k)}{q-k} \geq 0\,, \quad h_i(n) = h_i(q+k-n)\,, \quad \text{for every } n=k+1,...,q-1.$$

For a given subset $A\subset\Lambda$, let C_k^A denote the set of real positive definite functions on the cyclic group $(\mathbb{Z}_k)^A$ (for k=1, C_1^A denotes the set of non negative functions on A). Let $\mathcal{F}_A$ be the set of real functions of σ_A which vanish unless $\sigma_i \leq k$ for all $i\in A$ and whose restriction to $(\mathbb{Z}_k)^A$ belongs to C_k^A. Similarly, let $\mathcal{G}_A$ be the set of real functions of σ_A which vanish unless $\sigma_i \geq k+1$ for all $i\in A$ and whose restriction to the remaining subset with the structure of $(\mathbb{Z}_{q-k})^A$ belongs to C_{q-k}^A. We notice that the two above conditions express that h_i may be written as a sum

$$h_i^{(1)} + h_i^{(2)} \qquad \text{where} \qquad h_i^{(1)} \in \mathcal{F}_{\{i\}} \qquad \text{and } h_i^{(2)} \in \mathcal{G}_{\{i\}}$$

Let us write $\quad \mathcal{F} = \sum_{A\subset\Lambda} \mathcal{F}_A\,, \qquad \mathcal{G} = \sum_{A\subset\Lambda} \mathcal{G}_A$

(that is, $\mathcal{F}$ is the set of functions f which may be written as $f = \sum_{A\subset\Lambda} f_A$ with $f_A \in \mathcal{F}_A$).

Let finally $\langle \cdot \rangle$ denote the expectation value with respect to the Gibbs measure.

The following theorem can be proved with the generalizations of Ginibre inequalities [8], [11], [12], [13], [14].

Theorem 3. *Under the above hypothesis the following inequalities hold:*

$$\langle f\, g \rangle \leq \langle f \rangle \langle g \rangle \qquad \text{if } f \in \mathcal{F},\ g \in \mathcal{G}$$
$$\langle f_1 f_2 \rangle \geq \langle f_1 \rangle \langle f_2 \rangle \qquad \text{if } f_1, f_2 \in \mathcal{F}$$
$$\langle g_1\, g_2 \rangle \geq \langle g_1 \rangle \langle g_2 \rangle \qquad \text{if } g_1, g_2 \in \mathcal{G}$$

REFERENCES

1. J. W. Gibbs, On the equilibrium of heterogeneous substances, *Trans.Conn. Acad.* Vol III pp 343-524 (1878). Asso in *The Collected Works of J. Willard Gibbs*, Vol. 1, Longmans, Green, New-York (1928) p.258.
2. R. J. Baxter, Potts model at the critical temperature, *J. Phys. C: Solid State Phys.* **6**: L445 (1973), Magnetization discontinuity of the two dimensional Pott model, *J. Phys. A: Math. Gen.* **15**: 3329 (1982)
3. R. Kotecky, and S. B. Shlosman, First order phase transitions in large entropy lattice models, *Commun. Math. Phys.* **83**: 493 (1982)
4. L. Laanait, A. Messager and J. Ruiz, Phases coexistence and surface tensions for the Potts model, *Commun. Math. Phys.* **105**: 527 (1986)
5. R. Kotecky, L. Laanait, A. Messager and J. Ruiz, The q-state Potts model in the standard Pirogov-Sinaï theory, *J.Stat.Phys. (1989)*
6. J. De Coninck, A. Messager, S. Miracle-Solé and J. Ruiz, A study of perfect wetting for Potts and Blume-Capel models with correlation inequalities, *J. Stat.Phys.* **52**: 45 (1988).
7. A. Messager, S. Miracle-Solé and J. Ruiz, and S. B. Shlosman,"Interface in Potts model 2: Antonov rule and rigidity of the order disorder interface" Preprint C.P.T. June 1989
8. F.Dunlop, L. Laanait , A. Messager, S.Miracle Sole, J.Ruiz, Multilayer wetting in "q state" models. preprint C.P.T. June 89.
9. C. E. Pfister, Translation invariant equilibrium states in ferromagnetic abelian systems, *Commun. Math. Phys.* **86**: 375 (1982)
10. R.J. Baxter, " Exactly Solved Models in Statistical Mechanics" Academic Press 1982

11. F. Dunlop, Correlation inequalities for multicomponent rotators, *Commun. Math. Phys.* **49**: 247 (1976)

12. H. Kunz, C. E. Pfister and P. A. Vuillermot, Correlation inequalities for some classical vector models, *Phys. Lett.* **54A**: 428 (1975), *J. Phys. A: Math. Gen.* **9**: 1673 (1976)

13. J. L. Lebowitz, G.H.S. and others inequalities, *Commun. Math. Phys.* **35**: 87 (1974)

14. J. Ginibre, General formulation of Griffiths inequalities, *Commun. Math. Phys.* **16**: 310 (1970)

RANDOM SURFACES IN STATISTICAL MECHANICS

J. Bricmont

Physique Théorique U.C.L. 2, chemin du Cyclotron B-1348
LOUVAIN-LA-NEUVE BELGIUM

1 Introduction

In this lecture, I shall discuss results obtained in collaboration with A. El Mellouki and J. Frohlich, [1] and with J.L. Lebowitz and C. Maes [2,3]. They are all related to models of surfaces. These models occur in a number of fields, ranging from interfaces in statistical mechanics to gauge theories and strings. We shall start with models of interfaces separating two phases in thermal equilibrium. For example, an interface separating the plus and the minus phases in the Ising model or an A and a B phase in a mixture (where more phases, C, D etc ... may be present). We shall limit ourselves to SOS or Gaussian approximations to the interface. These approximations are expected to reproduce most qualitative properties of the real interface defined directly in terms of the Gibbs distribution of the original model.

The interface is modelled by a height variable, ϕ_x, one at each site x of a lattice $\mathbb{Z}^d$, which takes real or integer values, depending on whether the interface can vary continuously or not. The Hamiltonian is

$$H_\Lambda = \sum_{<xy>\cap\Lambda\neq\phi} |\phi_x - \phi_y|^\alpha \tag{1}$$

where for $\alpha = 1$ the model is of SOS type, and for $\alpha = 2$ it is called Gaussian. The dependence of the models on the values of α does not seem to be very important, but whether ϕ is discrete or continuous does matter. The boundary conditions ϕ_x, $x \notin \Lambda$, are fixed in (1).

The partition function is

$$Z_\Lambda = \sum_{\phi_x \in \mathbb{Z}, x\in\Lambda} exp(-\beta H_\Lambda) \tag{2}$$

for discrete ϕ, and

$$Z_\Lambda = \int_{-\infty}^{+\infty} \Pi_{x\in\Lambda} d\phi_x exp(-\beta H_\Lambda) \tag{3}$$

for continuous ϕ. Observe that β can be scaled away in the continuous model. If we scale β in the discrete model, and let β go to zero, one gets the continuous ϕ model. Also, notice that the continuous Gaussian model ($\alpha = 2$) is a good reference model since it is explicitly solvable.

These models are used for the description of one interface between two phases. The consideration of more than two phases leads one to introduce models where several interfaces interact. Before discussing these, let us review what is known about the phase diagram of the models with one interface.

The order parameter is the variance of the height : $< \phi_0^2 >_\Lambda$. The expectation value is taken with respect to the Gibbs ensemble with boundary conditions $\phi_x = 0$ $x \notin \Lambda$ (by translation invariance in ϕ space, any other fixed height will do). The question is whether $< \phi_0^2 >_\Lambda$ goes to infinity or not in the thermodynamic limit. If it does, the ϕ-translation invariance of the model, broken by the boundary conditions, is restored in that limit. Physically, it means that there is no stable interface between the two coexisting phases. By explicit computation, one shows that, in the continuous Gaussian case,

$$< \phi_0^2 >_\Lambda \simeq \begin{cases} |\Lambda| & d=1 \\ \log|\Lambda| & d=2 \\ < \infty & d \geq 3 \end{cases} \tag{4}$$

For the discrete models, in d=1,

$$< \phi_0^2 >_\Lambda \simeq |\Lambda| \tag{5}$$

so the model is in the high temperature phase for all β, which is to be expected in one dimension.

For 3 or more dimensions, it is always in the low temperature phase [4] :

$$< \phi_0^2 >_\Lambda < \infty \tag{6}$$

d=2 is the critical dimension : for T small $< \phi_0^2 > < \infty$ ([5]), while, for T large, $< \phi_0^2 > \simeq \log|\Lambda|$ ([6]). This last result, called the roughening transition, is the only hard result among those stated above.

2 Interacting surfaces

In [7] the following model of interacting surfaces was introduced in connection with the commensurate-incommensurate transition : consider a gas of surfaces, $\phi^i = (\phi_x^i)_{x\in} \mathbb{Z}^d$, $i = 1, \cdots, n(L)$ lying on top of each other, in $\mathbb{Z}^{d+1}$, each one having a Hamiltonian $H_\Lambda(\phi^i)$ given by (1) and with partition function in Λ, $|\Lambda| = (2L+1)^d$:

$$Z_\Lambda(\rho) = \sum_{\phi}^{\leq} exp[-\beta \sum_{i=1}^{n(L)} H_\Lambda(\phi^i)] \tag{7}$$

where the sum runs over $-L \leq \phi_x^1 \leq \phi_x^2 \leq \cdots \leq \phi_x^{n(L)} \leq L$.

One of the questions is to find how the free energy of the model depends on the density $\rho = \frac{n(L)}{2L+1}$ of the gas of surfaces. It appears that, except for $d = 1$ [10], it is very hard to get any exact information on this model. In particular, an interesting open question is to develop a kind of Mayer expansion for this gas of surfaces, at low densities, in $d = 2$ or more.

By making a further approximation, i.e. by considering every other surface as flat and rigid, one arrives at the model of one surface fluctuating between two rigid walls. The partition function is :

$$Z_\Lambda(\ell) = \sum_{\substack{|\phi_x| \leq \ell \\ \phi_x \in \mathbb{Z}}} exp(-\beta H(\phi)) \tag{8}$$

where H is given by (1). The sum is replaced by $\int_{-\ell}^{+\ell} \Pi_x d\phi_x$ for continuous ϕ's. ℓ here corresponds to ρ^{-1} in (7).

For the latter model, one can get precise estimates on the dependence of the free energy on the separation ℓ between the flat walls. These estimates are related to similar bounds on the correlation length of this model. While the latter is infinite for $\ell = \infty$, the cutoff $|\phi| \leq \ell$ produces a finite correlation length, and therefore the limit of low densities here is a sort of approach to a critical point (but a Gaussian one). This explains in part why setting up a Mayer expansion for these models is not straightforward. In [8] a type of mean field estimate was obtained for the divergence of the correlation length in this model :

$$\xi(\ell) \leq exp(c\ell^2) \tag{9}$$

We obtained in [1] the sharper, dimension dependent, bounds :

$$\xi(\ell) \simeq \left\{ \begin{array}{ll} l^2 & d = 1 \\ exp(c\ell) & d = 2 \\ exp(c\ell^2) & d \geq 3 \end{array} \right\} \tag{10}$$

The $d = 2$ result can be interpreted as a kind of "logarithmic" correction to mean field theory, the latter holding in $d \geq 3$. Here this logarithmic correction is rather easy to prove. Using (10), one can show, for the free energy, that :

$$|\Psi(\ell) - \Psi(\infty)| \simeq \left\{ \begin{array}{ll} l^{-2} & d = 1 \\ exp(-c\ell) & d = 2 \\ exp(-c\ell^2) & d = 3 \end{array} \right\} \tag{11}$$

One obtains also the rate at which $< \phi_0^2 >$ converges to its limit (which is infinite in $d = 1, 2$), see [1].

For the discrete models, with a cutoff $|\phi| \leq \ell$, one gets estimates similar to (11) on the rate of convergence of the free energy when the cutoff is removed. However in this discrete case, it is quite interesting to analyse the phase diagram of the model. First of all, for $\ell = \infty$, we have an infinity of phases whenever $< \phi_0^2 >< \infty$ (which occurs in $d \geq 3$ or in $d = 2$ at low temperatures, see Sect. 1). This just means that we can choose different boundary conditions in (1) leading to different values of $< \phi_0 >$. And this in turn expresses the fact that the symmetry of the Hamiltonian, given by $\phi_x \to \phi_x + a$ is spontaneously broken whenever $< \phi_0^2 >< \infty$.

For any $\ell < \infty$ odd, there is no symmetry left (for ℓ even, one has the $\phi \to -\phi$ symmetry), and there is no a priori reason to expect that one can adjust the boundary conditions so as to produce any desired values of $< \phi_0 >$. Indeed, for the continuous ϕ model, one easily shows that the presence of a finite correlation length implies that there is a unique phase, i.e. all boundary conditions lead to the same expectation values for all function of the field (see [1]).

For discrete ϕ's, the situation is less clear, and one may ask whether the model has some kind of phase transition. One can prove the following results : At low temperatures there is a unique phase, whose typical configurations consist of a "sea" of 0 spins with small islands of spins taking other values. However, there is no energy term in the Hamiltonian favouring the 0 configuration with respect to the other ones. The reason why only the 0 ground state survives at nonzero temperatures is purely entropic. Indeed, one easily computes that it has more low energy excitations than the other ground states. It is however not straightforward to prove that this entropic effect produces a unique phase at low temperatures. Usually, thanks to the Peierls'argument, one can deduce, at low temperatures, fairly general consequences based on energy considerations. But here we need the Pirogov-Sinai theory [9], which is a far-reaching extension of the Peierls'argument.

However this method is intrinsically limited to low temperatures and one may wonder whether some kind of phase transition occurs in this model at higher temperatures. All we can prove about this is that the susceptibility is bounded :

$$\Xi = \sum_{x \in \mathbb{Z}^d} < \phi_0 \phi_x > (\ell) < \infty (all d, \beta, \ell)$$

This makes a second-order transition rather unlikely, but not necessarily a first-order one. Also, there certainly exists a percolation transition in this model. Namely, at low temperatures, there is an infinite sea of 0 spins and the sites where $\phi_x \neq 0$ do not percolate. But, for ℓ not too small, there will be an infinite cluster of sites where $\phi_x \neq 0$ at high enough temperatures. This does not mean at all that there must exist a thermodynamic transition, defined by the coexistence of different phases and by singularities in the free energy.

3 Wetting and layering

The previous analysis of one surface fluctuating between two rigid walls has interesting consequences for the wetting problem where one considers one surface constrained to fluctuate above one wall. The partition function of this model is

$$Z_\Lambda = \sum_{\substack{\phi_x \in \mathbb{Z} \\ \phi_x \geq 0}} exp(-\beta H_\Lambda) \tag{12}$$

The surface ϕ represents the height of a droplet of one phase lying on a fixed substrate ("the wall"). This phase coexist with another one, which lies above the surface. The main issue here is the height of ϕ in the thermodynamic limit. If it goes to infinity, one says that the wall is (completely) wetted by the phase lying on it. In this simple model, it is easy to see that complete wetting always occurs for ϕ continuous and for discrete ϕ at low temperatures. Indeed this is just the same entropic effect as discussed in the previous section. The surface goes away from the wall in order to have more room to fluctuate (we stress that this means more room to fluctuate *towards the wall*; if we modify the model so as to forbid in the surface downwards spikes going towards the wall, no wetting will occur). Another way of saying this is to remark that the surface above a wall will not lift itself at a lower height above the wall than if it was constrained to remain below height 2ℓ. But then, we are back to the model discussed previously, where the surface settles in the middle, i.e. at height ℓ in the present notations. Since ℓ is arbitrary, $< \phi_0 > \rightarrow \infty$.

However, we must emphasize that the rate at which the height diverges as the thermodynamic limit is approached is extremely slow, i.e. as a power of the logarithm of the size of the box.

This phenomena of entropic repulsion has been shown to occur in greater generality by Lebowitz and Maes [12] : they consider any perturbation of the harmonic crystal (Hamiltonian (1) with $\alpha = 2$), which is not constant and which is monotone increasing in ϕ. This can be called a soft wall, a hard one being the one we considered :

$$H = \sum_{<xy>} (\phi_x - \phi_y)^2 - h \sum_x f(\phi_x), \quad h > 0, f \text{increasing}$$

They show that, with this potential, that

$$< \phi >_\Lambda \rightarrow +\infty \; as \; \Lambda \rightarrow \infty \tag{13}$$

This in turn motivates the following result in percolation theory, which illustrates the geometry of the Gaussian random field [2]. Consider the random set :

$$E_h(\phi) = \{x \in \mathbb{Z}^d : \phi_x \geq h\}$$

Does this set contain an infinite connected component? It is relatively easy to see that it does for $h \leq 0$. If it stops having one for large values of h, then we say that there is a percolation transition in that model (analogous to other percolation transitions). While such a transition is to be anticipated on intuitive grounds, it is not easy to prove via standard arguments, because the harmonic crystal has such strong correlations. The key fact in the proof is the instability of the harmonic crystal as expressed by (13). Basically, the reason why there is no infinite connected component in $E_h(\phi)$ for h large enough is that, if there was one, then the fields would have to be essentially everywhere at height h, which has probability zero. However the full argument is more delicate, but this is a basic input into it, see [2].

To obtain a wetting transition we must consider a less trivial model than (12). If we add to the Hamiltonian a term attracting the surface towards the wall, for example of the form

$$H_\Lambda = \sum_{<xy>} |\phi_x - \phi_y|^\alpha + \lambda \sum_x V(\phi_x) \tag{14}$$

where $V(\phi) = -exp(-k\phi), \lambda > 0$ (we consider here discrete ϕ's), then it is easy to show, via a Peierls'argument, that for any $\lambda > 0$ the surface is bound to the wall and the droplet is microscopic for β large. It is interesting to know how the surface unbinds itself from the wall i.e. how the height $< \phi_0 >$ diverges when λ increases. Using a rather detailed analysis based on the Pirogov-Sinai theory, one can show [17] that the unbinding of the surface goes through an infinite sequence of transitions-the layering transitions.

This means that one can find an infinite sequence of different curves, whose equations are approximately given by (4.10) in [1], in between which the surface is predominantly at height n, in the thermodynamic limit. The approximate equations are obtained by minimizing an approximate free energy given by an energy term, coming from the second term in (14), and an entropic term given by the spikes which are allowed at height n but not at height n-1, due to the constraint that the surface remains above the wall ($\phi \geq 0$).

4 Two random surfaces

A natural situation where wetting phenomena occur is when we have a coexistence of three or more phases and we consider a phase separation between two phases. It may happen that a third phase sets in between the two phases and wets the interface. A simple model where this should occur consists of two random interfaces forced to lie one on top of the other, with partition function given by :

$$Z_{\Lambda} = \sum_{\phi^1_x \geq \phi^2_x} exp[-\beta(H_{\Lambda}(\phi^1) + H_{\Lambda}(\phi^2)] \tag{15}$$

If the surface ϕ^2 is set equal to zero, we recover the previous model. So, one expect that the difference $\phi^1 - \phi^2$ of the heights will diverge in the thermodynamic limit. However, this is not so easy to prove, even using the Pirogov-Sinai theory, and is still an open problem. This just shows how little technology we have to handle statistical mechanics problems for extended objects, or, equivalently, "gases" with nonlocal constraints. Similarly, one may expect that, if we add an attractive potential between the two walls, a sequence of layering transitions will appear.

A related problem is to find criteria for the occurrence of this wetting phenomena in given models. So, assume we have three coexisting phases A,B,C; Can one tell whether phase C will wet an A-B interface? In principle, this question can be answered in a way similar to the question of which bulk phases will coexist at a given point in a phase diagram.

Namely, for the latter question, one associates to each potential phase a truncated free energy (which phases might occur is decided, for example, at low temperatures, if we know the ground states of the system: then we associate a phase to each ground state, by including the set of configurations which coincide with this ground states at most sites and differ from it only on small islands). Then we determine which phase(s) yield this minimum (truncated) free energy. The phases of the model are then given by those on which the minimum is reached. The true thermodynamic free energy of the model will then be well approximated by that minimum. All these ideas were made precise by Slawny [11] in the context of the Pirogov-Sinai theory.

Coming back to interfaces and wetting, we have tried to propose a similar scheme in [3], where, instead of the free energy, one has to compute the surface tension associated to different interfaces, i.e. an A-B interface, or successively an A-C and a C-B interfaces. Of course, we need some perturbative regime in order to be able to compute these approximate surface tensions.

Our results concern two models : First, the Blume-Capel model, which is just model (8) with $\ell=1$, but considered as a bulk model and not as a surface one. We argue that the zero phase wets a plus minus interface at low temperatures. The other model is the Potts model, where we use a large q expansion. We find that the disordered phase should wet any interface between two ordered phases, at the point where q+1 phases coexist. Neither of those results are rigorous. Further results on the wetting transition in these models can be found in [13-16].

References

1 J. Bricmont, A.El Mellouki, J. Fröhlich, J. Stat. Phys. **42**, 743 (1986).

2 J. Bricmont, J.L. Lebowitz, C. Maes, J. Stat. Phys., **48**, 1249 (1987).

3 J. Bricmont, J.L. Lebowitz, J. Stat. Phys., **46**, 1015 (1987).

4 J. Bricmont, J.R. Fontaine, J.L. Lebowitz, J. Stat. Phys. **29**, 193 (1982).

5 G. Gallavotti, A. Martin-Löf, S. Miracle-Sole, in Lecture Notes in Physics, **20**; (Springer, Berlin 1973).

6 J. Fröhlich, T. Spencer, Commun. Math. Phys. **81**, 527 (1981).

7 M. Fisher, D. Fisher, Phys. Rev. **B25**, 3192 (1982).

8 O. Mac Bryan, T. Spencer, Commun. Math. Phys. **53**, 299 (1977).

9 Ya. Sinaï, Theory of Phase Transitions : Rigorous Results (Pergamon-Press, New-York, 1982).

10 D.B. Abraham, in Phase Transition and Critical Phenomena, Vol. **11**, C. Domb, J.L. Lebowitz eds. (Academic Press, New York, 1988).

11 J. Slawny in Phase Transitions and Critical Phenomena, Vol.**11** C. Domb, J.L. Lebowitz eds. (Academic Press, New York, 1988).

12 J.L. Lebowitz, C. Maes, J. Stat. Phys. **46**, 39 (1987).

13 L. Laanait, A. Messager, J. Ruiz, Commun. Math. Phys. **105**, 572 (1986).

14 J. De Coninck, A. Messager, S. Miracle-Sole, J. Ruiz, J. Stat. Phys. **52**, 45 (1988).

15 P. Duxbury, J. Yeomans, J. Phys **A18** L983 (1985).

16 L. Armistead, J. Yeomans, J. Phys. **A20** 5635 (1987).

17 A. El Mellouki, Surfaces aléatoires, mouillage et transitions de couche. Thèse Louvain-La-Neuve 1987.

THE INFLUENCE OF BULK DISORDER ON WETTING PHENOMENA IN TWO DIMENSIONAL SYSTEMS

Th. M. Nieuwenhuizen

Institut für Theoretische Physik A RWTH, Templergraben 55, 5100 Aachen, Western Germany

Abstract: Disorder may modify the critical behaviour of two-dimensional wetting and depinning transitions in various ways. It is shown that first order transitions may be driven second order, that critical exponents of continuous transitions may be changed, that reentrant wetting may appear, and that correlated disorder may drive a continuous transition first order.

1 Introduction

Wetting phenomena are becoming more and more understood. A rich variety of interesting aspects has emerged from theoretical studies. Use has been made of several tools, ranging from exact solutions of 2-d Ising models [1] or 2-d Solid-on-Solid models [2], via functional renormalization group equations [3] and effective Landau-Ginzburg Hamiltonians [4] to mean field approaches [5]. Here we have referred to some of the many contributions in the specific fields. A much more complete discussion was given in a recent review by Dietrich [6].

Wetting may also take place in disordered systems. The simplest situation is that disorder is present in the substrate. A discussion of this problem has been given by H. Orland in a previous contribution to these proceedings [7]. Disorder in the bulk in three dimensions could describe, e.g., wetting by an interface located in a polymer mixture. Bulk disorder in two dimensions could describe a monolayer adsorbed on a random substrate, and 2-d wetting phenomena are assumed to occur inside this monolayer. Unfortunately, there are no general methods to solve systems with disorder. Even in one dimension, the way to derive, e.g., the exact density of states of a tight binding system with Cauchy distributed site disorder (Lloyd model [8]) is rather different from the way how solutions are obtained for diluted exponential distributions [9].

We shall consider the Restricted Solid-on-Solid (RSOS) limit of a two dimensional Ising model where spins are up at height $z = +\infty$. For considering wetting of a substrate we assume that spins are down at $z = 0$. For depinning of an interface from a line in the bulk (at $z = 0$) we will assume that spins are down at $z = -\infty$. In the RSOS model the interface separating the up and down phases is described by a height function z_i, where i is a label of the substrate sites. The interface may make single steps up or down ($z_{i+1} = z_i \pm 1$) or stay constant ($z_{i+1} = z_i$). It is known that the RSOS model has the same critical behaviour as the underlying, more complicated Ising model.

The Hamiltonian that we shall consider was essentially introduced by Huse and Henley [10].

$$H = J\sum_i |z_{i+1} - z_i| - \sum_i u_i\delta_{z_i,0} - \sum_i v_{i,z_i} \tag{1}$$

Here J is the renormalized bending energy, $u_i \equiv v_{i,0}$ the gain in energy due to the substrate and a positive (negative) v_{i,z_i} expresses the gain (loss) of energy if the interface has a height z_i at site i. The energies u_i and v_{i,z_i} come from breaking bonds of the underlying Ising model perpendicular to the substrate line $z = 0$. The bending energy J comes from breaking parallel bonds. Without loss of generality it has been assumed that the latter are non-random. The Hamiltonian (1) leads to a tridiagonal transfermatrix

$$(T_i)_{z,z'} = e^{\beta J|z-z'|}e^{\beta u_i\delta_{z,0}+\beta v_{i,z}} \tag{2}$$

where $z' - z = 0$ or ± 1 because of the RSOS limit. The partition sum may be defined as

$$Z_N = (T_1T_2\cdots T_N)_{00} \tag{3}$$

for a system of length N in the direction of the substrate. The surface free energy per substrate site is given by

$$-\beta F = lim_{N\to\beta}N^{-1}lnZ \tag{4}$$

It is instructive to consider wetting near a wall ($z_i \geq 0$) in the pure system, where $u_i = u$ and $v_i = 0$ [2a,b]. Here all transfer matrices are identical and one needs its right eigenvectors. The one with the largest eigenvalue is ψ_z = const. exp $(-z\mu)$. Its eigenvalue follows as $\lambda = 1 + 2e^{-\beta J}cosh\mu$ from bulk sites and as $\lambda = e^{\beta u}(1 + e^{-\beta J-\mu})$ from the substrate site. The parameter μ can be solved from these two equations. There are also plane wave solutions, $\psi_z \sim sinq(L - z)$, where L is the extension of the system in the direction perpendicular to the substrate. These plane waves have eigenvalues $\lambda = 1 + 2e^{-\beta J}cosq$ with $0 \leq q \leq \pi$ and q following from a quantization condition, derived in the same way as μ was determined above. For large L, q becomes uniformly distributed. For low enough temperatures the largest eigenvalue comes from the localized mode $\psi_z \sim exp(-z\mu)$. The wetting transition takes place when μ goes to zero. This happens at a temperature $T_w = 1/\beta_w$ given by

$$1 + 2exp(-\beta_w J) = exp(\beta_w u)(1 + exp(-\beta_w J)) \tag{5}$$

Close to the transition μ vanishes linearly, $\mu \sim t \equiv (T_w - T)/T_w$. This implies that the critical exponent ψ, related to the divergence of the average height, i.e. $< z >\sim t^{-\psi}$, and the exponent $\nu_\perp$, related to the correlation length perpendicular to the substrate, $\xi_\perp = \{< z^2 > - < z >^2\}^{\frac{1}{2}} \sim t^{-\nu_\perp}$, both take the value $\psi = \nu_\perp = 1$. Furthermore, the free energy behaves as $-\beta f = ln\lambda_{max} = ln(1+2exp(-\beta J)) + exp(-\beta J)\mu^2$, as compared to $-\beta f = ln(1 + 2exp(-\beta J))$ for delocalized modes with $q = 0$. It follows that the specific heat makes a jump at the wetting transition, implying a specific heat critical exponent $\alpha = 0$. The parallel correlation length $\xi_{||}$ describes the decay of correlations in the direction of the substrate, viz. $< z_i - z_j > \sim exp(-|i-j|/\xi_{||})$. It is equal to the inverse of the energy gap between the localized and delocalized states :

$$\xi_{||} = (\Delta E)^{-1} = e^{\beta J}/\mu^2 \tag{6}$$

This implies a parallel correlation length exponent $\nu_{||} = 2$. Another quantity of interest is the wandering exponent ς. For an unbound interface, wandering over a distance $L_{||}$ in the parallel direction, it is defined by

$$L_{\perp} \equiv \{< (z_{L_{||}} - z_0)^2 >\}^{\frac{1}{2}} \sim L_{||}^{\varsigma} \tag{7}$$

For a bound interface the same relations holds when the correlation lengths are inserted. This gives the general relation $\varsigma = \nu_{\perp}/\nu_{||}$. In the present situation where disorder is absent it follows that $\varsigma = \frac{1}{2}$, which is just the random walk value.

If the interface depins from a line in the middle of the bulk (at $z = 0$ with $-L \le z \le L$ and $L \to \infty$) the critical behaviour remains unchanged. The wetting temperature will now follow from the equation exp $(\beta_w u) = 1$, i.e. $T_w = \infty$. In the Ising model T_w should be identified with the bulk critical temperature, which is shifted towards infinity in the RSOS limit. A more useful control parameter, therefore, is the line attraction energy u. When this vanishes, the interface has no reason to remain localized and depins. The critical value is $u_c = 0$, rather than $u_c > 0$ for wetting of a wall, because there is no loss of entropy when an interface is pinned near a line in the bulk.

We shall now consider various forms of disorder in the Hamiltonian (1). The situation where only substrate disorder is present (u_i independent random variables with common distribution $\rho_L(u)$ and $v_{i,z}$ vanishing) was treated by H. Orland in another contribution to these proceedings [7].

2 Interface that can be pinned near a wall and near an impurity line in the bulk.

Here we consider a system where an interface may be pinned by a substrate, which we choose to be at height $z = -L$, and by a random line potential at $z = 0$. The system ranges from $z = -L$ to $z = +L$ and L goes to infinity in the thermodynamic limit. We assume no further bulk potentials, so $v_{i,z} = 0$ except for $v_{i,-L} = u_s$. The line at $z = 0$ is disordered and the distribution of each u_i is $\rho_L(u_i)$. In the thermodynamic limit the problem separates in solving the pure wetting transition near a wall (see above) and the depinning transition from a disordered line in the bulk. The latter problem is close to the topic discussed by Orland [7]. In particular, the critical point of the RSOS model is known exactly, and follows from the equation [11].

$$exp\{\beta U_e(\beta)\} \equiv \int \rho_L(u) e^{\beta u} du = 1 \tag{8}$$

This condition for depinning comes from the annealed system, where the partition sum is averaged over disorder. It also gives the transition point of the quenched system, where the free energy is averaged over disorder. The depinning transition from the line in the bulk occurs when the effective line potential $U_e(\beta)$ vanishes. We now follow Ref. [12] and consider the situation where $u_i = u_A = 1$ with probability p and $u_i = u_R = -4$ with probability $1-p$. We further assume that $J = 1$ and $u_s = 0.3$. If the line is attractive on the average, $pu_A + (1-p)u_R > 0$, the interface wants to be pinned by the line at all temperatures (unless, for low temperatures, the substrate is more attractive; this does not happen in our case). This situation occurs in Figure 1 for $p_d < p < 1$. When the line is repulsive on the average, the interface can be pinned to it at low temperatures because of

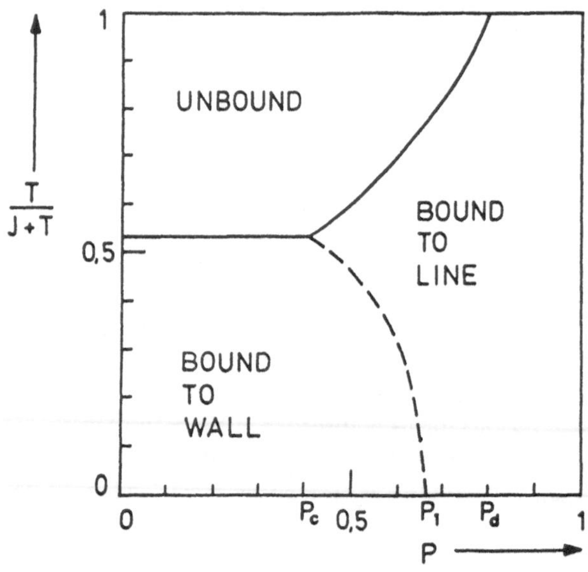

Fig. 1. Phase diagram for a system where the interface can be pinned near a wall or to a disordered line in the bulk. Full lines : second order transition lines, the broken line is a first order line.

gain of energy in attractive regions; but it will be delocalized for high temperatures. This is the only transition which occurs for $p_1 < p < p_d$, and the nature of the transition is the same as for depinning from a random substrate [7,11] : in particular the jump in the specific heat has a 1/log singularity due to disorder. For $p < p_1$ in Figure 1 the line will be too repulsive to pin the interface at zero temperature. Hence the interface will be located near the substrate for low temperatures. For $p_c < p < p_1$ the free energy of the substrate bound state will become equal to the one of the line bound state at a certain temperature $T_1 > 0$. Here a first order transition takes place where the interface jumps from the substrate to the impurity line in the bulk. At a larger temperature T_2 will depin continuously from the impurity line and be unbound. For $p \to p_c$, T_1 and T_2 coalesce. For $p < p_c$ the impurity line is too repulsive to play any role. Then there is only the previously discussed continuous depinning transition from the pure substrate.

The present model shows an explicit example where, due to disorder, a second order transition line (for $p < p_c$) splits up in a first order and another second order line in the phase diagram. If the wall attraction is stronger (larger u_s) it may happen that the pure system at $p = 1$ has a wall bound phase for low temperature and a line bound phase at higher temperature. The phase diagram will differ from Figure 1 in the sense that the broken line now goes down less steep and ends at a finite height at p=1. In going from right to left in the new phase diagram, this first order transition line will coalesce with a second order line to end in another second order line. Also this example shows explicitly that large enough disorder may qualitatively change the phase diagram.

3 Wetting of a substrate in the presence of bulk disorder

The system with Hamiltonian (1), fixed $u_i \equiv u$ and bulk disorder ($v_{i,z}$ independent random variables for $z \geq 1$) was studied by Kardar [13]. Assuming Gaussian disorder, $< v >\equiv 0, < v^2 >= \Delta_v$ he introduced the replica method and calculated $< Z_N^n >$. This function can be calculated for n=1,2,3,.... The results are assumed to be also valid when the integer n is sent to zero, where $< Z_N^n >= 1 + n < lnZ_N > +0(n^2)$ gives the free energy. The replicated partition function involves n interfaces which have a site independent transfermatrix because translational invariance is restored by performing the average over disorder at all sites. The replicated Hamiltonian is

$$H_n = J\sum_\alpha |z_\alpha - z'_\alpha| - u\sum_\alpha \delta_{z_\alpha,0} - \frac{1}{2}\beta\Delta_v \sum_{\alpha,\beta} \delta_{z_\alpha,z_\beta}(1 - \delta_{z_\alpha,0}) \tag{9}$$

where α and β are replica indices $(1 \leq \alpha, \beta \leq n)$. Instead of disorder, there is a two particle interaction between the interfaces, and the additional parameter n has to be taken small at an appropriate moment. The eigenvalue equation related to (9), i.e., $e^{-\beta H_n}\Psi = \Lambda\Psi$, can be studied in the continuum limit where $e^{-\beta J}$, βu and $\beta^2\Delta_v$ are small parameters. One inserts $\Lambda = [1 + 2e^{-\beta J}]^n(1 - E)$ and obtains a Schrödinger equation $H_c\Psi = E\Psi$ with

$$H_c = -D\sum_\alpha \frac{\partial^2}{\partial z_\alpha^2} + \sum_\alpha U(z_\alpha) - \Delta_v \sum_{\alpha<\beta} \delta(z_\alpha - z_\beta) \tag{10}$$

where $D = 1/(e^{\beta J} + 2)$ is the diffusion coefficient and $U(z)$ is a square well potential, $U = \infty$ for $z < 0$, $U = -\overline{u}$ for $0 < z < 1$ and $U = 0$ for $z > 1$. The value of $\overline{u}$ is chosen such that the bound state of $U(z)$ has the same decay exp $(-\mu z)$ as the bound state of the discrete one particle Hamiltonian (1) with $v_{i,z} = 0$ and u_i replaced by $u - \beta\Delta_v/2$. This shift comes from the $\alpha = \beta$ terms in (9). It is solely a consequence of the fact that the wall is not disordered. In going from (9) to (10) two-particle terms at $z = 0$ are neglected. In the problem with only substrate disorder such a term only leads to logarithmic corrections to the critical behavior, whereas it does not shift the wetting temperature [7,11].

The ground state of the many body Hamiltonian

$$H_{MB} = D\{-\sum_\alpha \frac{\partial^2}{\partial z_\alpha^2} - 4\kappa \sum_{\alpha<\beta} \delta(z_\alpha - z_\beta)\} \tag{11}$$

was derived by Mc Guire [14]

$$\Psi = exp(-\kappa \sum_{\alpha<\beta} |z_\alpha - z_\beta|) \tag{12}$$

Kardar argued that the approximate ground state of (10) is

$$\Psi = exp\{-(\mu + n\kappa - \kappa)\sum_\alpha z_\alpha - \kappa \sum_{\alpha<\beta} |z_\alpha - z_\beta|\} \tag{13}$$

where $\kappa = \beta^2 \Delta_v/(4D)$ has the dimension 1/length and is induced by disorder. Eq.(13) is a solution in the sector where all particles are outside the well (all $z_\alpha > 1$) and it has the correct behavior when one of them gets close to the well. The mismatch in the other sectors is expected to be negligible near the depinning transition. As is seen from (13), unbinding takes place when $\mu + n\kappa - \kappa$ vanishes. For $n \to 0$ this occurs at $\mu = \mu_c \equiv \kappa$, rather than at $\mu_c = 0$ in the pure model. This shift shows that bulk disorder has the tendency to delocalize the interface, and it is half as large as the shift already occurring because the substrate is not disordered. The cause of this new shift is the presence of energetically favorable regions far away in the bulk. If, rather than the bulk, the substrate is disordered, such favorable regions will be found at the substrate and the interface will be stronger bound than without disorder. From (13) the quenched free energy can be calculated.

$$\begin{aligned} -\beta F &= ln(1 + 2e^{-\beta J}) - lim_{n\to 0}(E/n) \\ &\simeq D\{2 + 2\kappa - \kappa^2/3 + (\mu - \kappa)^2\} \end{aligned} \tag{14}$$

It has a quadratic behavior near the transition $\mu \to \kappa$, so also in the disordered model the specific heat makes a jump at the wetting point. Whether there is a 1/log correction due to surface terms omitted in (10) is not known. The average height $< z >$ and the correlation length $\xi_\perp$ can also be calculated from (13), and are given in section 5. It follows that the critical exponents $\psi = \nu_\perp = 2$ differ from their pure values. This is not unexpected, since the Harris criterion says that bulk disorder is relevant for dimensions $d > 5/3$ [10]. Initial numerical verifications of these new exponents by multiplication of the random transfer matrices (3) were unsuccessful [13]. However, by changing the wall coupling parameter u, rather than the strength of disorder Δ_v or the temperature, this behavior has been clearly observed. In the next section figures are presented for a closely related situation.

4 Depinning from a line in the bulk

We now consider the same problem as in the previous section, but assume that the z-values range from $-\infty$ to $+\infty$, and that the interface can be pinned by the line $z = 0$. In this model the continuum Hamiltonian (10) can be simplified by setting $U(z) = -2D\mu\delta(z)$. However, no solution of this many particle problem has been reported.

The critical point of this system is simpler to obtain [15]. In the annealed model it occurs when the line is, on the average, just as attractive as the bulk, i.e. when

$$\int \rho_L(u) e^{\beta u} du = \int \rho_B(v) e^{\beta v} dv \tag{15}$$

where ρ_L is the distribution of the line energies u_i and ρ_B the one of the random bulk energies $v_{i,z}(-\infty < z < \infty; z \neq 0)$. This expression is a generalization of (8), where only the line was disordered. In the replicated disordered system the difference between bulk terms and line terms will only involve two, three,$\cdots$, particle terms at the point given by (15). For the related problem with solely substrate disorder [11] it is known that such terms do not shift the critical point, and the same is assumed to happen here. Eq. (15) allows for reentrant wetting transitions. It may happen that the interface is delocalized for temperatures $0 < T < T_1$ and for $T > T_2$, but is pinned by the line for $T_1 < T < T_2$. This suggests that also the model of previous section may exhibit reentrant wetting behavior if both bulk and substrate are disordered.

In Figures 2 and 3 we present numerical calculations of the free energy and the inverse of the average of the absolute height. We study a typical case, where $T = 1, e^{-J} = 0.3$ and the bulk potentials take the values $\pm 1/\sqrt{2}$ with equal probabilities. For both quantities a quadratic behavior near the unbinding transition at $u = u_c$ is clearly observed. This supports the expectation that the critical behavior of the models of sections 4,5 and 6 is the same.

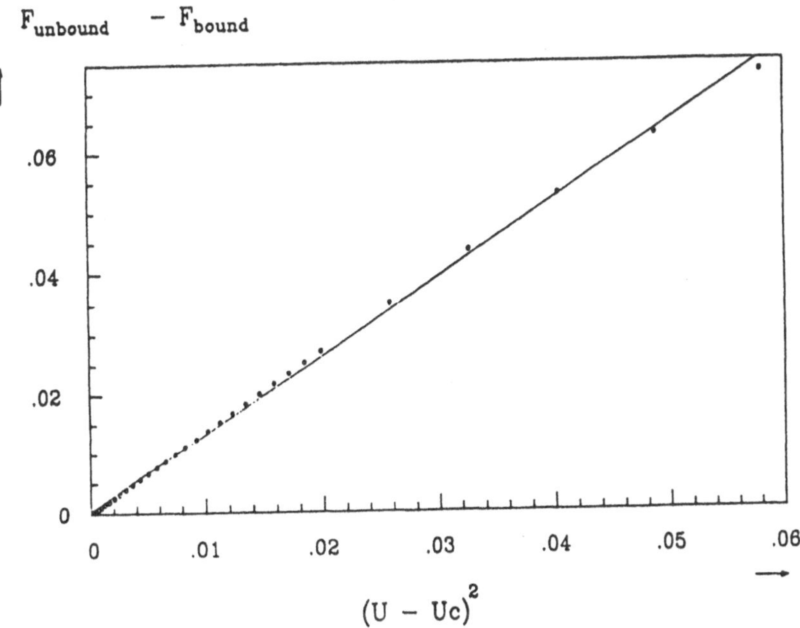

Fig. 2. The free energy near the unbinding transition as function of the strength of the line potential.

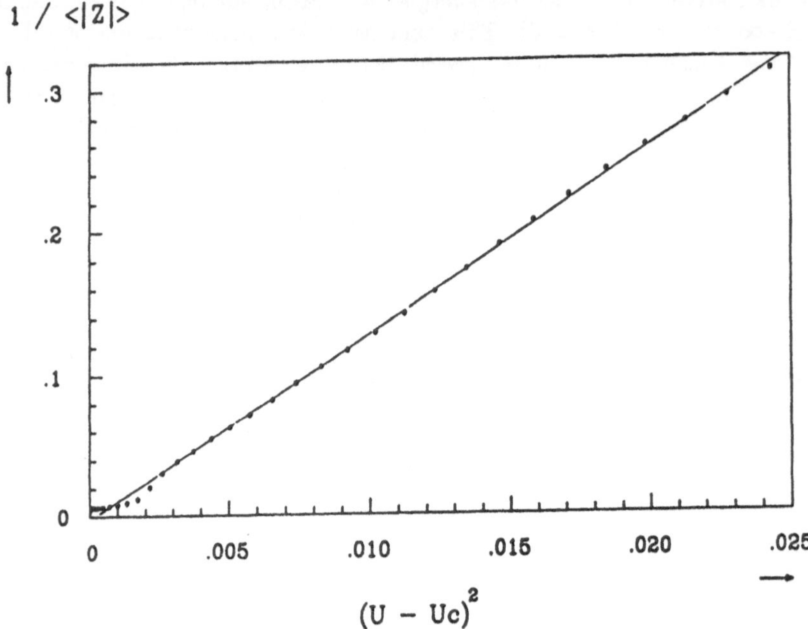

Fig. 3. The inverse of the average of the absolute height near the unbinding transition. Note its quadratic behavior.

5 Depinning from a line in a bulk with symmetric disorder

Here we consider the same system as in section 4, but now we assume that disorder is symmetric with respect to the line at $z = 0$, i.e. $v_{i,z} = v_{i,-z} = v_{i,|z|}$. Such a system can be realized, in principle, by cutting a crystal and glueing the parts side-by-side. The continuum Hamiltonian for this system is [15]

$$H = D\{-\sum_{\alpha} \frac{\partial^2}{\partial z_\alpha^2} - 2\mu \sum_{\alpha} \delta(z_\alpha) - 4\kappa \sum_{\alpha<\beta} \delta(|z_\alpha| - |z_\beta|)\} \tag{16}$$

The absolute values arise from the assumed symmetry. Restricting ourselves to symmetric solutions and going to new variables $y_\alpha = |z_\alpha|$, (16) takes the form (11), together with the boundary conditions

$$\frac{\partial}{\partial y_\alpha}\Psi + \mu\Psi = 0 \qquad (y_\alpha \to 0, \alpha = 1, 2, \cdots, n) \tag{17}$$

which is usual for delta functions. The exact solution of this problem is given by (13), with $y_\alpha = |z_\alpha|$ replacing z_α. We conclude that the model considered here confirms the results of Kardar, discussed in section 3.

The problem posed by equations (11) and (17) can be solved exactly by the Bethe ansatz. The usual boundary condition that Ψ vanishes when one of the y_α vanishes, can simply be generalized to satisfy eq. (17). As a result, the energy gap can be calculated.

The idea is to split up the n particles in different groups which behave as free molecules (apart from the first one, which is localized near $z = 0$). In this sense (12) represents one molecule with n particles, and (13) a molecule with n particles bound to the line $z = 0$.

The energy of the optimal partition is

$$\Delta E = \frac{D}{6\kappa}(\mu - \kappa)^3 \tag{18}$$

It follows that the parallel correlation length $\xi_{\|} = 1/\Delta E$ has a critical exponent $\nu_{\|} = 3$. From the previous result $\nu_{\perp} = 2$ the wandering exponent follows as $\varsigma = 2/3$. This result was first conjectured from numerical data by Huse and Henley [10] and then derived by means of the Burgers equation [16]. A simpler argument was given by Kardar [13]. He observed that in the continuum limit an unbound interface has a replicated free energy $ln < Z_N^n >= N(an + bn^3)$. Here the term linear in n determines the average free energy. The quadratic term in n is its second cumulant, which is seen to vanish, i.e., it is of lower order $(N^{\frac{2}{3}})$ in N. The third cumulant shows that fluctuations in lnZ_N are of order $N^{\frac{1}{3}}$, which implies $\varsigma = 2/3$ by a scaling argument. Our approach, described above, is the first one where $\varsigma = 2/3$ has been derived by studying a depinning transition.

We can also determine some prefactors. It is found from (13) that

$$\overline{< |z| >} = \frac{\kappa}{(\mu - \kappa)^2} \tag{19}$$

and

$$\overline{< z^2 > - < z >^2} = \overline{< z^2 >} = \frac{4\kappa^2}{(\mu - \kappa)^4} \tag{20}$$

which leads to

$$\xi_{\perp} = \overline{< z^2 >}^{\frac{1}{2}} = (\frac{2\kappa D^2}{9})^{\frac{1}{3}} \xi_{\|}^{\frac{2}{3}} \tag{21}$$

This result can be compared with a prediction of Nattermann [17]. He studies the unbound situation, where $\xi_{\perp}$ stands for the width of the interface, and $\xi_{\|} \equiv L$ for its length in the "time" direction. Inserting $D = T/\Gamma$ in (21), where Γ is the stiffness coefficient (normalized to unity in [17]), we find full agreement with eq. (8) of ref. [17]. Moreover, we have a precise value for an undetermined numerical factor. It should be noted, however, that the two problems are not fully equivalent; the numerical factor in Nattermann's model may not be the same as in ours.

Surprisingly, we also find

$$\overline{< z^2 > - < |z| >^2} = \frac{2\kappa}{(\mu - \kappa)^3} \tag{22}$$

which implies that the interface is relatively sharp near the phase transition. This is caused by the fact that the paralel correlation length diverges faster than the perpendicular correlation length. Nevertheless, the average absolute height $< |z_i| >$ itself exhibits large fluctuations in the direction of the pinning line, as follows from the fact that, to leading order, $\overline{< |z| >^2} = 4\overline{< |z| >}^2$, due to (19), (20) and (22).

6 Wetting when disorder in the bulk is strongly correlated

We finally consider a system [18] where the wall is not disordered ($u_i = u$ in eq.(1)) and where disorder in the bulk is fully correlated in the direction of the wall ($v_{i,z} = v_z$ for $z = 1, 2, \cdots$). The bulk properties of the underlying Ising model were studied by Mc Coy and Wu [19]; see [20] for a recent, exact solution. Because the transfer matrices of the RSOS model are already site independent, it is not necessary to introduce replicas. Instead one searches the eigenvectors of the tridiagonal matrix (2). This corresponds to a one-dimensional disordered system, for which all eigenfunctions are localized exponentially with probability one.

To be specific, we consider binary disorder, where v_z vanishes with probability q and is repulsive ($v_z = -\overline{v}$) with probability $p = 1 - q$. For small enough temperature, or for deep enough wall potential, the groundstate will be the exponentially localized substrate bound state. Its eigenvalue is denoted by λ_μ. In the disordered system, this state will be exponentially localized near the wall for any temperature or strength of wall potential. There is also a continuum of (localized) bulk states, with eigenvalues in the range $(1 - 2exp(-\beta J))exp(-\beta\overline{v}) \leq \lambda \leq 1 + 2exp(-\beta J) \equiv \lambda_B$. In particular, a state with eigenvalue $\lambda = 1 + 2\gamma cos\epsilon$ close to λ_B is localized in a strip of width $n \simeq \pi/\epsilon$ where all v_z vanish, marked by two lines where $v_z = -\overline{v}$. Narrower strips have smaller eigenvalues; wider ones allow larger eigenvalues. This phenomena is nothing but the Lifshitz band edge singularity in the density of states of disordered systems [20]. The wetting transition occurs when λ_μ crosses λ_B, at a temperature $T = T_w$. Due to disorder the surface bound state remains localized. The bulk state, however, is infinitely separated. Hence the transition is first order, and there is non-vanishing latent heat. Nevertheless, the transition has a linearly diverging parallel correlation length $\xi_{||} = (\lambda_\mu - \lambda_B)^{-1} \sim (T_w - T)^{-1}$. This is typical for first order transitions in two dimensions, whereas in higher dimensions $\xi_{||}$ usually stays finite at a first order transition.

The wetting temperature of the correlated system varies from sample to sample. The reason is that the free energy of the surface bound state strongly depends on the realization of disorder close to the substrate. As opposed to the situation in previous sections, there is no self averaging of disorder in the direction of the wall. The bulk state, on the other hand, always has the free energy $f_B = -Tln(\lambda_B)$ when the system is infinitely large. Figure 4 shows the density of possible wetting temperatures for a specific case ($q = 0.75, \overline{v} = J = 0.25, u = 0.4$).

It is also interesting to consider the system away from coexistence. Then a term $h \sum_i z_i$ has to be added to the Hamiltonian (1). It can be derived that [18].

$$< z > \sim h^{-1} |lnh|^{-3} \tag{23}$$

for small h. This differs from the mean field value $< z > \sim |lnh|$ and the behaviour $< z > \sim h^{-1/3}$ in an ordered two dimensional system. The behavior of (20) will not be monotonous. The interface will be located in lanes without disorder of width proportional to $|lnh|$ and make rather sharp transitions according to (20). Only for $h \to 0$ these transitions become real steps. So there are no real phase transitions (layering transitions or prewetting line) in the system off coexistence. This is due to the two dimensional nature of the system, and related to the presence of a diverging correlation length.

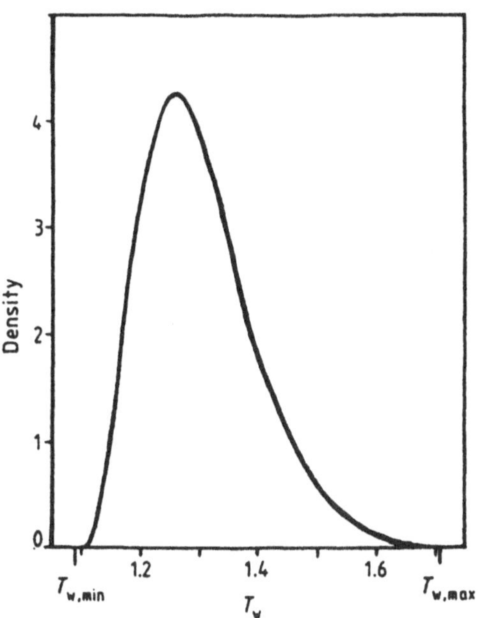

Fig. 4. Distribution of possible wetting temperatures, related to different realizations of disorder for different samples of the ensemble of systems with correlated binary disorder.

7 Summary and outlook

We have discussed several models for wetting phenomena, where disorder is present in the bulk couplings. Many interesting aspects have appeared. In section 2 disorder is only present on a line, and if it is strong, it can modify the phase diagram. Sections 3, 4 and 5 treat closely connected models with small bulk disorder, where the critical behavior differs from the one in the pure system. Finally, in section 6 we consider a model with correlated disorder, where the wetting temperature varies from sample to sample.

Our discussion has been limited to two dimensional wetting transitions, which can occur inside adsorbed monolayers. Unfortunately experiments on such systems have not yet been performed. From a theoretical side, studies in this direction seem very promising, and new results will probably be discovered. Many interesting aspects have not been solved, such as the influence of long range interactions and of random bulk fields.

Acknowledgements

It is a pleasure to thank Stefan Gryglewski and Bernard Tsai for many discussions and for their help in preparing the manuscript.

References

1 D.B. Abraham in Phase transitions and Critical Phenomena Vol. 10, C. Domb and J.L. Lebowitz, eds. (Academic, NY, 1986).

2 J.M.J. van Leeuwen and H.J. Hilhorst, Physica 107A (1981) 319 and S.T. Chui and J.D. Weeks, Phys. Rev. B23 (1981) 2438.

3 R. Lipowsky and M.E. Fisher, Phys. Rev. Lett. 56 (1986) 472.

4 C. Ebner and W.F. Saam, Phys. Rev. Lett. 58 (1987) 587 and J.O. Indekeu, Nucl. Phys. B (Proc. Suppl.) 5A (1988) 168.

5 R. Pandit, M. Schick and M. Wortis, Phys. Rev. B26 (1982) 5112.

6 S. Dietrich, in Phase Transitions and Critical Phenomena, vol. 12, C. Domb and J.L. Lebowitz, eds. (Academic, NY, 1988).

7 H. Orland, these proceedings.

8 P. Lloyd, J. Phys. C2 (1969) 1717.

9 Th. M. Nieuwenhuizen, Phys. Lett. 103A (1984) 333.

10 D.A. Huse and C.L. Henley, Phys. Rev. Lett. 54 (1985) 2708.

11 G. Forgacs, J.M. Luck, Th. M. Nieuwenhuizen and H. Orland, Phys. Rev. Lett. 57 (1986) 2184; J. Stat. Phys. 51 (1988) 29.

12 G. Forgacs and Th. M. Nieuwenhuizen, J. Phys. A21 (1988) 3871.

13 M. Kardar, Phys. Rev. Lett. 55 (1985) 2235; Nucl. Phys. B290 (FS20) (1987) 582.

14 J.B. Mc Guire, J. Math. Phys. 5 (1964) 622.

15 Th. M. Nieuwenhuizen, to appear

16 D.A. Huse, C.L. Henley and D. Fisher, Phys. Rev. Lett. 55 (1985) 2924.

17 T. Nattermann, Europhys. Lett 4 (1987) 1241.

18 Th. M. Nieuwenhuizen, J. Phys. A21 (1988) L567

19 B.M. Mc Coy and T.T. Wu, The two dimensional Ising Model, (Harvard, Cambridge, 1973).

20 Th. M. Nieuwenhuizen and H. Orland, Phys. Rev. B, to appear (1989).